Instructor's Guide to accompany

A Laboratory Textbook of
Anatomy & Physiology

SEVENTH EDITION

Anne B. Donnersberger

Anne E. Lesak

Moraine Valley Community College

JONES AND BARTLETT PUBLISHERS
Sudbury, Massachusetts
BOSTON   TORONTO   LONDON   SINGAPORE

World Headquarters
Jones and Bartlett Publishers
40 Tall Pine Drive
Sudbury, MA 01776
978-443-5000
info@jbpub.com
www.jbpub.com

Jones and Bartlett Publishers Canada
2100 Bloor St. West
Suite 6-272
Toronto, ON M6S 5A5
CANADA

Jones and Bartlett Publishers International
Barb House, Barb Mews
London W6 7PA
UK

Copyright © 2000 by Jones and Bartlett Publishers, Inc.

ABOUT THE COVER
Illustration of the muscles of the human male body, anterior view: © J. Daugherty, Science Source/Photo Researchers, Inc.
Chest and leg radiographs: © PhotoDisc, Inc.
Photograph of foot: © PhotoDisc, Inc.
Photomicrograph simple columnar epithelium of human colon: © Turtox, Inc.

All rights reserved. No part of the material protected by this copyright notice may be reproduced or utilized in any form, electronic or mechanical, including photocopying, recording, or by any information storage and retrieval system, without written permission from the copyright owner.

ISBN 0-7637-1053-9

Printed in the United States of America

03 02 01 00 99   10 9 8 7 6 5 4 3 2 1

To The Instructor

This *Guide* is for instructors who are using *A Laboratory Textbook of Anatomy and Physiology*, Seventh Edition. It is designed to facilitate laboratory preparation for the instructor, teaching assistant, or laboratory coordinator.

The Guide contains (1) answers to the Self-Test, Discussion Questions, and Case Studies contained in the textbook; (2) a list of supplies, equipment, and solutions needed for each exercise in the textbook for a class of 32 students working individually, in pairs, or in groups; (3) a list of major vendors; and (4) unlabeled illustrations from the textbook.

CD-ROM activities are suggested in selected exercises. They are marked in both the *Laboratory Textbook* and the *Instructor's Guide* with the icon in the margin. You can find them on the CD-ROM entitled *The Virtual Physiology Lab* (ISBN 0-697-37994-9) available from WCB/McGraw-Hill and Cypris Publishing. Call 1-800-262-4729 to order.

We wish to acknowledge the assistance of Glenn McDonald in the preparation of this *Guide*. We hope you will find it to be helpful.

Anne B. Donnersberger

Anne E. Lesak

Contents

Answers to Questions and Case Studies
- Unit 1   Medical Terminology   3
- Unit 2   The Microscope   5
- Unit 3   Cells   6
- Unit 4   Tissues   8
- Unit 5   The Skeletal System   10
- Unit 6   The Musculature System   14
- Unit 7   The Nervous System   17
- Unit 8   The Special Senses   20
- Unit 9   The Blood, Lymphatic, and Cardiovascular Systems   22
- Unit 10   The Respiratory System   27
- Unit 11   The Digestive System   29
- Unit 12   The Urinary System   32
- Unit 13   Acid-Base Balance   34
- Unit 14   The Reproductive System   36
- Unit 15   The Endocrine System   38

List of Materials and Supplies
- Unit 1   Medical Terminology   43
- Unit 2   The Microscope   44
- Unit 3   Cells   45
- Unit 4   Tissues   48
- Unit 5   The Skeletal System   50
- Unit 6   The Musculature System   52
- Unit 7   The Nervous System   55
- Unit 8   The Special Senses   58
- Unit 9   The Blood, Lymphatic, and Cardiovascular Systems   61
- Unit 10   The Respiratory System   66
- Unit 11   The Digestive System   68
- Unit 12   The Urinary System   72
- Unit 13   Acid-Base Balance   75
- Unit 14   The Reproductive System   77
- Unit 15   The Endocrine System   80

Major Vendors   83

Unlabeled Figures   87

Answers to Questions and Case Studies

UNIT 1

Medical Terminology

ANSWERS TO QUESTIONS WITHIN EXERCISES

1.1. Coronal (or frontal)

1.2. Mediastinum: heart, trachea, esophagus, aortic arch
Dorsal superior cavity: brain

1.3. Left hypochondriac: spleen
Right iliac: appendix

1.4. Left upper quadrant: spleen, stomach
Right lower quadrant: appendix, cecum of large intestine

1.5. Variable responses, including:
    ad: adduction
    con: concave, concomitant, concurrent
    ex: excoriation, exocrine, excise
    hyper: hyperactivity, hyperextension, hyperglycemia
    super: superficial, superinfection

1.6. Variable responses, including:
    gastro: gastroenteritis, gastroenterology
    hepato: hepatocyte, hepatomegaly
    derm: dermatitis, dermabrasion

1.7. -ectomy: to remove
-ostomy: to create an opening
-otomy: to cut into

1.8. Variable responses, including:
    -algia: myalgia
    -itis: arthritis, laryngitis, gastritis
    -ology: cardiology, neurology

ANSWERS TO SELF-TEST QUESTIONS

(1) h	(2) f	(3) a	(4) c	(5) a	(6) c	(7) b
(8) d	(9) d	(10) d	(11) c	(12) b	(13) a	(14) b
(15) c	(16) d	(17) b	(18) a	(19) c	(20) c	(21) a
(22) d	(23) d	(24) c	(25) d			

DISCUSSION QUESTIONS

26. a) Language is universal because of the adoption of a standardized anatomical vocabulary.
b) The language of anatomy describes the body in precise terms.
c) Anatomical terms cannot be misunderstood.

27. a) The language of anatomy can be confusing to the beginning student.
b) Many of the terms have synonyms.

28. Variable responses.

Unit 1

ANSWERS TO CASE STUDIES

1. Lumbar
   Kidneys
   Blunt impact to kidneys, causing bleeding
   No, there is probably renal involvement.

2. The receptionist submitted to the insurance company records stating that there was a need for x-rays of the ileum, not ilium. The ileum is that last portion of the small intestine, soft tissues and not visible through x-ray. On the other hand, Jenny's problem was a fracture of her ilium bone, a dense tissue and visible through x-ray imaging. The error in spelling caused the insurance company's refusal to pay the bill because incorrect information was conveyed to them.

UNIT 2

The Microscope

ANSWERS TO QUESTIONS WITHIN EXERCISES

2.1. Eyepiece and objectives
2.2. Low or scanning
2.3. Backwards and upside down
2.4. Oil immersion
2.5. So the light from the base will shine through the objective, enabling the slide to be viewed

ANSWERS TO SELF-TEST QUESTIONS

(1) e	(2) d	(3) c	(4) b	(5) a	(6) d	(7) d
(8) b	(9) d	(10) b	(11) a	(12) c	(13) c	(14) d
(15) b	(16) c	(17) b	(18) d	(19) a	(20) d	(21) b
(22) b	(23) c	(24) a	(25) d			

DISCUSSION QUESTIONS

26. Raise the condenser or increase the aperture of the diaphragm.

27. The immediate focusing of an image under a microscope when changing from one magnification to another without adjusting the coarse or fine adjustment knobs

28. It would distort the image when using the 43X objective (high power). The diaphragm should be left completely open on most microscopes. Closing the iris diaphragm while using this objective is undesirable because it reduces resolution by diminishing the numerical aperture of the condenser.

ANSWER TO CASE STUDY

Rhonda should locate the iris diaphragm under the stage. By slowly moving the lever at the side of the diaphragm, she can permit more or less light to enter the microscope, thereby illuminating the field to desired level of clarity.

UNIT 3

Cells

ANSWERS TO QUESTIONS WITHIN EXERCISES

UNIT 3A - CELL STRUCTURE

3A.1. Nucleus, nucleolus, cytoplasm (cytosol), plasma membrane (or cell membrane)

3A.2. Nucleus: contains DNA in chromosomes for cell division and protein synthesis
Nucleolus: contains ribosomal RNA, which functions in protein synthesis
Cytoplasm: contains water, dissolved particles, organelles, and cytoskeleton
Plasma membrane: forms outer boundary of cell, selectively allows molecules to enter and exit cell

3A.3. No, they are too small to be seen with a light microscope.

3A.4. 46

3A.5. 46

UNIT 3B - CELL PHYSIOLOGY

3B.1. Variable responses - need a membrane, a hypertonic solution, and a hypotonic solution

3B.2. Osmosis

ANSWERS TO SELF-TEST QUESTIONS

(1) a (2) d (3) c (4) a (5) b (6) c (7) b
(8) c (9) a (10) c (11) c (12) a (13) b (14) b
(15) d (16) d (17) c (18) c (19) c (20) c (21) c
(22) b (23) c (24) d (25) c

DISCUSSION QUESTIONS

26. a) **chromatin** - (DNA) contains hereditary information of the cell.
   b) **nucleolus** - stores ribosomal RNA.
   c) **centrioles** - involved in spindle formation and movement of chromosomes during cell division.
   d) **plasma membrane** - external membrane of cell which regulates interchanges of molecules in and out of the cell.
   e) **endoplasmic reticulum** - site of carbohydrate, lipid, and protein synthesis and is involved in transport of macromolecules throughout the cell.
   f) **ribosomes** - sites of protein synthesis.
   g) **Golgi apparatus** - collects and packages products of cell synthesis to be stored until they are secreted.
   h) **lysosomes** - contain hydrolytic enzymes which function in splitting protein, mucopolysaccharide, and glycogen molecules.
   i) **mitochondria** - contain oxidative enzymes for breaking down products of glucose metabolism into water, CO_2, and ATP.
   j) **nuclear membrane** - regulates interchanges of molecules in and out of the nucleus.

27. a) interphase
    b) telophase
    c) metaphase
    d) prophase
    e) prophase, metaphase, anaphase, early telophase
    f) interphase
28. The size of the molecules

ANSWER TO CASE STUDY

No, Paul failed to recognize that mature red blood cells have no nuclei. He may have mistaken the clearer central region of the erythrocyte as a nucleus. The simple squamous epithelial oral cells and the stratified squamous epithelial cells from his skin epidermis have a nucleus, cytoplasm, and cell membrane.

UNIT 4

Tissues

ANSWERS TO QUESTIONS WITHIN EXERCISES

4.1.   Secrete mucus
4.2.   Connective
4.3.   An opening through which a fluid flows
4.4.   Haversian (central) canal
4.5.   To harden, or replace with bone
4.6.   Fat
4.7.   Involuntary
4.8.   Variable responses
4.9.   Variable responses
4.10. Variable responses
4.11. Variable responses

ANSWERS TO SELF-TEST QUESTIONS

(1) d   (2) a   (3) a   (4) c   (5) b   (6) a   (7) b
(8) c   (9) a   (10) b (11) d (12) d (13) d (14) c
(15) d (16) a (17) d (18) b (19) a (20) c (21) b
(22) b (23) c (24) d (25) a

DISCUSSION QUESTIONS

26. Student Drawings
27. a) stratum basale
    b) stratum corneum, stratum lucidum, part of stratum granulosum
    c) stratum corneum, stratum lucidum, part of stratum granulosum
28. a) cuboidal epithelium
    b) traditional epithelium
    c) pseudostratified, ciliated, columnar epithelium, or hyaline cartilage
    d) smooth muscle
    e) simple squamous epithelium and loose connective (areolar)
    f) striated (skeletal) muscle
    g) dense, regular connective tissue
    h) hyaline cartilage, pseudostratified ciliated columnar epithelium
    i) simple columnar epithelium
    j) adipose tissue

ANSWER TO CASE STUDY

John was partially correct in his identification of and functions of tissues in the ureter x-section. John erred, as do many beginning histology students, in not recognizing transitional epithelium lining the lumen. He should have observed more closely the polygonal cells in the mid-tissue region and the balloon cells toward the lumen itself. Squamous cells would have appeared flattened throughout the tissue. John had mistakenly identified transitional epithelium for stratified squamous epithelium. Transitional provides the necessary stretching required of the ureter to provide for urine flow to the bladder. Smooth muscle responsible for involuntary contraction and movement of urine is found in the muscularis regions of the ureter x-section. The serosal cover (adventitia) of the ureter is dense, white, fibrous connective tissue and provides a protective covering. John correctly identified these two tissues.

UNIT 5

Skeletal System

ANSWERS TO QUESTIONS WITHIN EXERCISES

UNIT 5A - SKELETAL ANATOMY

5A.1. The end with the large knob on it (the head)

5A.2. To prevent friction between the bones of a joint

5A.3. Upper division; there are more carpals than tarsals

5A.4. A tuberosity is larger

5A.5. Variable responses

5A.6. Jugular: opening through which lateral sinus and cranial nerves IX, X, XI pass
Carotid: opening through which internal carotid artery passes

5A.7. Stylomastoid foramen: opening through which facial nerve passes
Jugular fossa: depression in which internal jugular vein lies

5A.8. Lacrimal groove: serves as tear ducts
Alveolar process of the maxilla: forms socket into which teeth are anchored

5A.9. Cervical: 7
Thoracic: 12
Lumbar: 5
Sacral: 5 (fused in adult)
Coccygeal: 3-5 (fused in adult)

5A.10. The spinal cord

5A.11. The inferior articular surface of T-10 vertebra

5A.12. T-12 has inferiorly pointing spinous process; inferior articular surfaces face laterally.
L-1 is larger and heavier with bladelike posterior projecting spinous process; inferior articular processes directed ventrally and laterally.

5A.13. Transverse processes contain a transverse foramen.

5A.14. It has no body.

5A.15. It forms a pivot point to turn the head.

5A.16. Atlas and C-3

5A.17. 12

5A.18. They need to bear more weight.

5A.19. The uniting of two or more bones

5A.20. Manubrium, gladiolus (or body), xiphoid

5A.21. Sternum and vertebrae

5A.22. Vertebrae

5A.23. Scapula, clavicle, humerus

5A.24. Anatomical neck is immediately inferior to head, surgical neck is inferior to greater and lesser tubercles (and is broken more frequently, hence requiring surgery)

5A.25. Radius

5A.26. Coronoid

5A.27. Olecranon

5A.28. Two

5A.29. Ischial tuberosity

5A.30. In the acetabulum

5A.31. Inner ankle: medial malleolus of tibia
Outer ankle: lateral malleolus of fibula

5A.32. Three

UNIT 5B - IDENTIFICATION OF JOINTS

5B.1. Anterior

5B.2. To accommodate brain growth

5B.3. Dense, fibrous connective tissue (regular)

5B.4. Spheroidal

ANSWERS TO SELF-TEST QUESTIONS

(1) d	(2) b	(3) c	(4) a	(5) a	(6) c	(7) d
(8) c	(9) b	(10) b	(11) a	(12) b	(13) b	(14) c
(15) a	(16) b	(17) b	(18) a	(19) b	(20) d	(21) d
(22) c	(23) c	(24) a	(25) b			

DISCUSSION QUESTIONS

26. a) **trochlea** - a pulley-shaped part of a bone; trochlea of humerus.
   b) **tubercle** - a small, rounded projection to which muscles attach; adductor tubercle of femur.
   c) **trochanter** - a large, irregularly shaped projection to which muscles attach; greater trochanter of femur.
   d) **spine** - a sharp, needle-like projection serves as a point for muscle attachment; anterior nasal spine.
   e) **process** - an outgrowth or projection of bone; zygomatic process.
   f) **fossa** - an indentation or depression in bone; jugular fossa.
   g) **foramen** - a hollow opening that serves as a passageway for nerves and blood vessels; jugular foramen.
   h) **condyle** - a rounded, convex projection; occipital condyle.
   i) **groove** - a narrow, slit-like depression which is not as deep as a fissure; intertubercular groove of humerus.

Unit 5

27.

Axial Skeleton	**Appendicular Skeleton**
Skull	Clavicle
Hyoid	Scapula
Vertebrae	Humerus
Sternum	Radius
Ribs	Ulna
	Carpals
	Metacarpals
	Phalanges
	Ossa coxae
	Femur
	Patella
	Tibia
	Fibula
	Tarsals
	Metatarsals
	Phalanges

28.
- a) **mandibular fossa** - forms socket for mandibular condyle.
- b) **jugular foramen** - serves as a passageway for lateral sinus and cranial nerves IX, X, & XI
- c) **carotid foramen** - serves as passageway for internal carotid artery
- d) **sella turcica** - surrounds pituitary gland
- e) **superior orbital fissure** - allows for passage of cranial nerves III, IV, & VI
- f) **foramen rotundum** - transmits maxillary branch of cranial nerve V
- g) **crista galli** - serves for attachment of meninges
- h) **foramen magnum** - serves as outlet for spinal cord from brain
- i) **superior articulating surface of a vertebra** - serves as point of attachment for superior vertebrae
- j) **odontoid process** - provides for pivot-type attachment of axis to atlas
- k) **rib tubercle** - attachment of ribs to vertebrae
- l) **acromion of scapula** - serves as point of articulation with clavicle
- m) **glenoid cavity** - serves as point of articulation of scapula with humerus
- n) **coronoid fossa of humerus** - serves as articulation with coronoid process of ulna
- o) **semilunar notch of ulna** - serves as articulation with trochlea of humerus
- p) **ischial tuberosity** - serves as support for body in sitting position
- q) **acetabulum** - serves as articulation of pubis with head of femur
- r) **trochlea of femur** - serves as articulation with patella and tibia
- s) **intercondylar eminence** - serves as articulation between tibia and patella
- t) **trochlea of humerus** - serves as an articulation with ulna and radius

Skeletal System

  u) **sternal end of clavicle** - the medial, rounded end which articulates with the sternum
  v) **costal cartilage of rib** - medially positioned cartilage articulating with sternum or other cartilage (costal cartilages of the 8th through 10th ribs attach to each other and then to costal cartilage of 7th rib)

ANSWERS TO CASE STUDY

1. Student Drawings
2. Because Janet is 21 years old, her epiphyseal plate is closed. Growth in long bones such as the femur is complete. Janet, therefore, has no need to be concerned.

UNIT 6

Muscular System

ANSWERS TO QUESTIONS WITHIN EXERCISES

UNIT 6A - HUMAN MUSCULATURE

6A.1.   Orbicularis oculi
6A.2.   Platysma
6A.3.   Scalenus
6A.4.   Inspiration
6A.5.   Expiration
6A.6.   Vertically
6A.7.   Horizontally
6A.8.   Superior to the spine
6A.9.   Inferior to the spine
6A.10. Biceps brachii
6A.11. Triceps brachii
6A.12. Triceps brachii
6A.13. Adducts the hand: flexor carpi radialis
        Extends the wrist and inserts at the base of the fifth metacarpal: extensor carpi ulnaris
6A.14. Gluteus maximus
6A.15. Gastrocnemius

UNIT 6B - CAT MUSCULATURE

6B.1.   Muscles will be more difficult to separate connective tissues will be more difficult to cut.
6B.2.   Masseter and digastric
6B.3.   Sternohyoid, stylohyoid, thyrohyoid
6B.4.   Pronator teres
6B.5.   Transversus abdominis
6B.6.   Pectoralis minor
6B.7.   Cat: three
        Human: one
6B.8.   Epitrochlearis
6B.9.   Biceps femoris
6B.10. Extends shank (foreleg)
6B.11. Gastrocnemius

ANSWERS TO SELF-TEST QUESTIONS

(1) d (2) c (3) b (4) a (5) b (6) b (7) a
(8) d (9) c (10) d (11) a (12) c (13) b (14) c
(15) a (16) d (17) b (18) d (19) d (20) d (21) a
(22) b (23) d (24) c (25) b

DISCUSSION QUESTIONS

26. (Student drawing).
27. a) biceps femoris, semimembranosus, semitendinosus
    b) rectus femoris, vastus lateralis, vastus medialis, vastus intermedius
28. a) **Pectoralis major** - larger than pectoralis minor in human, smaller than pectoralis minor in cat
    b) **Pectoralis minor** - smaller than pectoralis major in human, larger than pectoralis major in cat
    c) **Pectoantebrachialis** - not present in human
    d) **Xiphihumeralis** - not present in human
    e) **Sternomastoid** - analogous to sternocleidomastoid muscle in human (two heads in human); one muscle in cat
    f) **Clavotrapezius, Acromiotrapezius, Spinotrapezius** - three muscles in cat, one muscle (trapezius) in human
    g) **Levator scapulae ventralis** - not present in human
    h) **Clavodeltoid, Acromiodeltoid, Spinodeltoid** - one muscle (deltoid) in human, three muscles in cat
    i) **Biceps brachii** - long head flexes arm and forearm in human, short head supinates hand in human; single head flexes in cat
    j) **Caudofemoralis** - not present in human
    k) **Gluteus maximus** - larger than gluteus medius muscle in humans, smaller than gluteus medius muscle in the cat
    l) **Sartorius** - narrow muscle in human, wide muscle in cat
    m) **Adductor femoris** - one muscle in cat, three muscles (adductor longus, adductor brevis, adductor magnus) in human
    n) **Epitrochlearis** - not present in human

ANSWERS TO CASE STUDY

1. Deltoids, biceps, levator scapulae, serratus anterior, trapezius, rhomboids, latissimus dorsi, and pectoralis minor and major
2. Connective tissue: humerus, cartilage, tendons, ligaments, fascia; skeletal muscle, brachial plexus, and nerves; axillary artery and its branches

3. Cartilage is avascular, so the extent of damage to this tissue influences the length of healing time. Pulled muscles and other connective tissues heal fairly fast in a normal, healthy 25-year-old male. Rupture of the vascular supply can cause serious bleeding into the joint cavity. Damage to the brachial plexus disrupts innervation to the entire upper extremity, including the fingers. Therefore, the extent of overall tissue damage is a crucial factor in repair and estimating total length of time to fully heal.

UNIT 7

Nervous System

ANSWERS TO QUESTIONS WITHIN EXERCISES

UNIT 7A - NERVOUS SYSTEM ANATOMY

7A.1. Variable responses, including cell body (nucleus, perikaryon), axon, dendrites, Nissl bodies
7A.2. In the cerebrum
7A.3. Between the gray and white matter of the cerebellum
7A.4. Variable responses, including posterior median sulcus, posterior median septum, white matter, posterior horns, anterior horns, gray matter, anterior median sulcus, lateral horns
7A.5. Sensory: receive input from body regions and perceive sensations
Motor: control skeletal muscle activities
7A.6. Motor coordination: cerebellum
Primary sensations: sensory areas
Internal homeostasis: hypothalamus
7A.7. Pons: contains apneustic and pneumotoxic respiratory centers
Medulla: contains respiratory, cardiac, vasomotor, and reflex centers
Hypothalamus: controls body temperature, appetite, fluid balance, and sexual drive
7A.8. I - Olfactory, and II - Optic

UNIT 7B - NERVOUS SYSTEM PHYSIOLOGY

7B.1. The motor branch of the vagus nerve is involuntary and innervates the pharynx and esophagus.
7B.2. No, there are individual differences.
7B.3. To protect the retina from damaging light rays
7B.4. To change the shape of the lens of the eye (becomes more convex) to focus on a near object
7B.5. To focus on a near object
7B.6. The heel moves upward rapidly
7B.7. Protection from foreign objects irritating the cornea
7B.8. Variable responses
7B.9. Variable responses
7B.10. Pupil should constrict when exposed to light
7B.11. Variable responses
7B.12. Variable responses
7B.13. Variable responses; should be able to feel sensations of touch, heat, and cold
7B.14. Variable responses
7B.15. Should be able to differentiate beween salt and sugar
7B.16. Variable responses
7B.17. Variable responses
7B.18. Variable responses
7B.19. Variable responses
7B.20. Variable responses

Unit 7

CHART FOR CRANIAL NERVES

Number	Name and Branches	Sensory or Motor	Function
I	olfactory	sensory	smell
II	optic	sensory	vision
III	oculomotor	motor	movement of eyeball; adjustment in size of lens and shape of pupil
IV	trochlear	motor	moves eyeball superiorly
V	trigeminal	sensory and motor	
	3 Divisions:		
	1) ophthalmic	sensory	general sensational impulses of pain, temperature, and pressure from optic orbit, face, and scalp
	2) maxillary	sensory	general sensational impulses from upper mouth and lips, teeth, and nasal cavities
	3) mandibular	motor and sensory (mixed)	chewing, general sensation from lower lips and teeth
VI	abducens	motor	movement of eyeball laterally
VII	facial	motor and sensory	facial expression, taste from anterior two-thirds of tongue
VIII	vestibulocochlear (acoustic/auditory)	sensory	
	2 Divisions:		
	1) vestibular	sensory	equilibrium
	2) acoustic	sensory	hearing
IX	glossopharyngeal	sensory and motor	taste from posterior two-thirds of tongue, salivation
X	vagus	sensory and motor	regulates secretions from gastric glands and pancreas; muscle sense of larynx and pharynx; respiratory, cardiac, and circulatory reflexes; slows heart rate; increases peristalsis
XI	accessory (spinal accessory)	motor	muscular action of pharynx and larynx; muscle action of sternocleidomastoid and trapezius; shares some functions with vagus nerve
XII	hypoglossal	motor	muscular action of tongue in speech, chewing, and swallowing

ANSWERS TO SELF-TEST QUESTIONS

(1) c   (2) d   (3) b   (4) c   (5) a   (6) b   (7) b
(8) a   (9) b   (10) b (11) a (12) b (13) c (14) a
(15) b (16) b (17) d (18) a (19) c (20) a (21) c
(22) c (23) d (24) a (25) b

DISCUSSION QUESTIONS

26. Mature neurons are not capable of undergoing mitosis because they have lost the ability to form spindle fibers.

27.
    a) **olfactory bulbs** - contain cell bodies of cranial nerve I; fibers from olfactory tract
    b) **pyramidal tracts** - fiber tracts of the cerebral peduncles on the anterior surface of the medulla, which connects the spinal cord and cerebellum to higher centers of the brain
    c) **optic chiasma** - formed by the crossing of the medial portion of the optic nerve from each eye; the nerve fibers then extend to the occipital lobe and brain stem as the optic tracts
    d) **thalamus** - serves as a relay station and integration center between cerebral cortex and spinal cord
    e) **hypothalamus** - the homeostatic center of ANS; functions in regulating appetite, body temperature, water balance, hormone secretion
    f) **cerebellum** - maintains posture, gait, and coordination
    g) **arbor vitae** - contains tracts which communicate with the midbrain, pons, and higher brain centers

28. A spinal reflex is one which is centered in the spinal cord, without the immediate involvement of the brain. The reflex arc depends not only on the afferent and efferent fibers, but on the central structure where the fibers are connected. Examples: knee jerk (patellar reflex), stretch reflex, ankle jerk (Achilles reflex).

29. The primary ions are sodium and potassium (Sodium-Potassium pump). Calcium is also needed for nerve impulse transmission.

30. Student's interpretations of frog responses to stimuli.

ANSWERS TO CASE STUDY

Mark could have suffered either a hemorrhage or a concussion. A concussion temporarily disrupts brain function when the brain moves too abruptly within the cranial vault. A hemorrhage is a more serious consideration because blood leaves vessels and builds up around the brain itself or around the meninges.

Serious damage to the posterior aspect of the skull might affect the cerebellum, which is an area of the brain that controls coordination. Visual regions of the brain are located in the occipital lobe of the cerebrum just above the cerebellum.

UNIT 8

Special Senses

ANSWERS TO QUESTIONS WITHIN EXERCISES

UNIT 8A - THE EYE AND VISION

8A.1. Rotates eyeball superiorly and laterally

8A.2. Rods: function in black and white (night) vision
Cones: function in color vision

8A.3. Variable responses

8A.4. Lens

8A.5. Medial (or interior, or anterior)

UNIT 8B - THE EAR, HEARING, AND EQUILIBRIUM

8B.1. External auditory (acoustic) meatus

8B.2. Stapes

8B.3. Scala media

8B.4. Superior and/or inferior and/or lateral semicircular canals

8B.5. Variable responses

8B.6. Laterally to the left and right

8B.7. Variable responses

UNIT 8C - OLFACTORY, TASTE, AND CUTANEOUS RECEPTORS

8C.1. Sour

8C.2. Variable responses

8C.3. Decreases taste

8C.4. "tasters" - variable responses
"nontasters" - variable responses

8C.5. Fewer nerve endings in wrist than palm

8C.6. Yes; variable responses

8C.7. Most sensitive: inner wrist
Least sensitive: sole

8C.8. Palm; variable responses

8C.9. Sole; variable responses

8C.10. Distal upper extremities more sensitive than lower extremities

8C.11. Palm

8C.12. Yes; variable responses

Special Senses

ANSWERS TO SELF-TEST QUESTIONS

(1) c   (2) d   (3) a   (4) b   (5) a   (6) b   (7) c
(8) d   (9) d   (10) a (11) b (12) a (13) c (14) d
(15) c (16) b (17) b (18) c (19) a (20) b (21) c
(22) d (23) a (24) a (25) d

DISCUSSION QUESTIONS

26. Circular and radial smooth muscle fibers of the iris

27. These tests differentiate between conduction and sensineural deafness and compare hearing in both ears. In the Weber test, a person with conduction deafness will hear vibrations in the poorer ear. A person with sensineural hearing loss will hear them in the better ear. A person with normal hearing will localize the sound as being in front of him. In the Rinne test, if there is a conductive loss, vibrations through the mastoid process will be louder than those heard by air conduction. If there is sensineural loss or normal hearing, vibrations through the air will be louder.

28. No, the degree of ability to taste is an inherited characteristic. Individuals vary greatly in their ability to recognize sweet, sour, bitter, and salty variations of taste. There are also some diseases in which taste is impaired. Olfactory sensation also plays a part in taste interpretation.

29. Density of corpuscles of Meissner in the papillary skin layer, numbers of free nerve endings around hair follicles and density of Pacinian (lamellated) corpuscles beneath the skin, as well as variations in stimulation influence variations in ability to sense touch.

30. Most commonly, colorblind people have a deficiency either in the red or green range. The second most common type of color blindness is in the blue-yellow range.

ANSWER TO CASE STUDY

Marva's problem is a thickening of her eardrum, a tissue between her outer and middle ear. Because of scar tissue, the flexibility of the ear drum has been compromised, and it does not vibrate in the normal manner. Consequently, the ear ossicles of the middle ear do not transmit the required intensity of vibration from the source waves that hit the stiffened tympanic membrane. The resultant stimulation of the cochlear nerve is reduced and the perception of sound lessened. Marva has, therefore, reduced hearing.

Her otologist is an M.D., ear specialist.

UNIT 9

The Blood, Lymphatic, and Cardiovascular Systems

ANSWERS TO QUESTIONS WITHIN EXERCISES

UNIT 9A - ANATOMY OF THE BLOOD, LYMPHATIC SYSTEM, BLOOD VESSELS, AND HEART

9A.1. They have a larger body mass.
9A.2. Neutrophils, lymphocytes, monocytes, eosinophils, basophils
9A.3. Nucleus
9A.4. So fluid balance can be maintained
9A.5. No
9A.6. Tonsil, spleen, thymus
9A.7. Thoracic and abdominal
9A.8. Left upper (and left lower to lesser extent)
9A.9. Yes; functions are assumed by bone marrow and liver
9A.10. Blood flowing through a vein is under less pressure.
9A.11. Single layer thick wall
9A.12. Blood pressure in venae cavae and pulmonary veins is too low to open them.
9A.13. Tricuspid: prevents blood from backing up into right atrium
Bicuspid (mitral): prevents blood from backing up into left atrium
9A.14. Blood in ventricles is under greater pressure.
9A.15. Pulmonary artery
9A.16. Aorta
9A.17. Left inferior lateral
9A.18. Left; it is under greater pressure
9A.19. Usually an advantage; allows heart chambers to expand and contract
9A.20. Left and right ventricles
9A.21. Two; tricuspid and bicuspid
9A.22. Two; pulmonary semilunar valve and aortic semilunar
9A.23. Yes; the pulmonary artery transports deoxygenated blood from the heart to the lungs
9A.24. Axillary: armpit
Brachial: arm
9A.25. Yes; the pulmonary veins transport oxygenated blood from the lungs to the left atrium of the heart
9A.26. One
9A.27. No
9A.28. Aorta → left subclavian artery → axillary artery → brachial artery → ulnar artery

The Blood, Lymphatic, and Cardiovascular Systems

9A.29. Posterior tibial vein → popliteal vein → deep femoral vein → femoral vein → external iliac vein → common iliac vein → inferior vena cava → right atrium

9A.30. (1) Ductus venosus
(2) Ductus arteriosus

UNIT 9B - HUMAN BLOOD PHYSIOLOGY

9B.1. Type O
9B.2. Type A blood contains B antibodies, which would agglutinate the red blood cells.
9B.3. Grams per 100 mL of blood
9B.4. Hemoglobin content of males is greater
9B.5. The cell membranes burst
9B.6. Variable responses (8 - 16 g/100 mL)
9B.7. Variable responses (hemoglobinometer is more specific)
9B.8. 84
9B.9. Radial artery
9B.10. Should be the same
9B.11. It would obstruct blood flow to the brain.
9B.12. Palpation would be the same or slightly lower.
9B.13. The aortic arch curves to the left.
9B.14. The ear canals are positioned anteriorly in the temporal bones.
9B.15. Variable responses (100/60 - 140/90)
9B.16. Variable responses
9B.17. Variable responses (auscultation probably lower)
9B.18. Hearing acuity; using incorrect cuff size; positioning of stethoscope on arm; background noise
9B.19. Positioning of microphone on arm; snugness of cuff around arm

UNIT 9C - CARDIAC MUSCLE PHYSIOLOGY

9C.1. 1 - d
2 - c
3 - a
4 - b

9C.2. In a healthy cardiovascular system, yes. If a cardiac or vessel problem, not all heart beats would be effectively transmitted through the blood vessels.

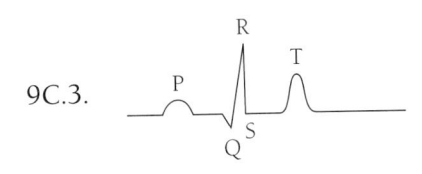

9C.3. *[ECG tracing with labeled P, Q, R, S, T waves]*

9C.4. Exercise: increases heart rate
Holding breath: decreases heart rate

24 Unit 9

9C.5. 1) Change in intrathoracic pressure
2) Change in oxygen concentration of blood
3) Change in pH of blood

9C.6. Increases it

9C.7. Emotions, exercise, metabolic rate, oxygen, and carbon dioxide content of blood

9C.8. Cardiac muscle has a slower action potential (it takes longer to contract).

9C.9. It slows the heart rate.

9C.10. No, it is excitatory (increases heart rate).

ANSWERS TO SELF-TEST QUESTIONS

(1) a   (2) c   (3) a   (4) a   (5) c   (6) b   (7) d
(8) c   (9) a   (10) a   (11) d   (12) c   (13) c   (14) d
(15) a   (16) c   (17) c   (18) a   (19) d   (20) d   (21) c
(22) d   (23) d   (24) a   (25) b

DISCUSSION QUESTIONS

26. **FLOW CHART - HUMAN BLOOD CLOTTING**

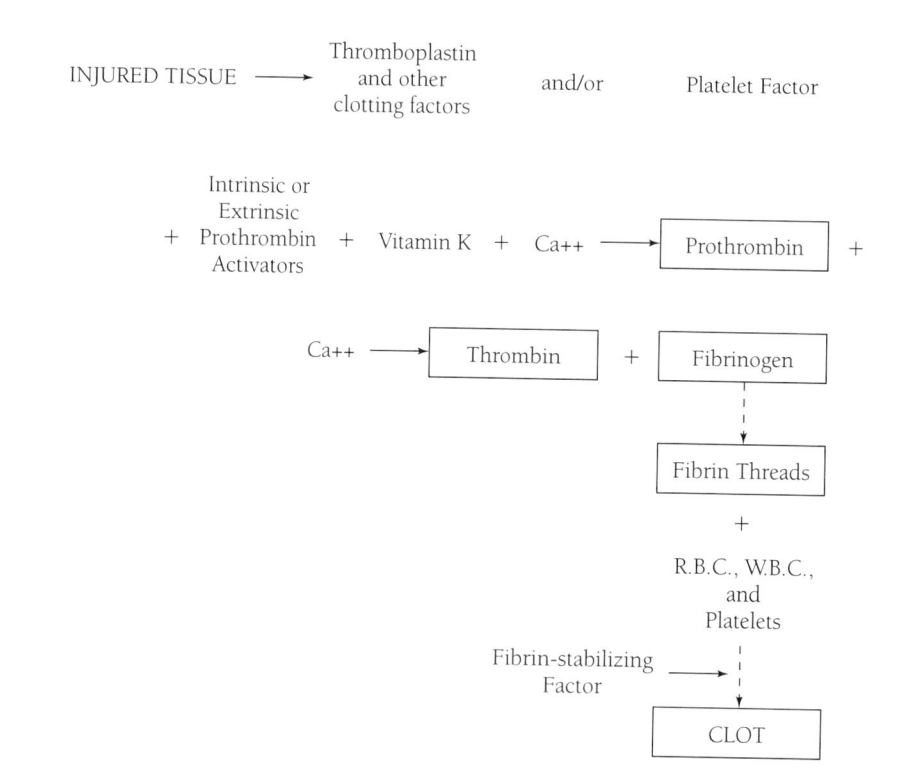

INJURED TISSUE → Thromboplastin and other clotting factors   and/or   Platelet Factor

+ Intrinsic or Extrinsic Prothrombin Activators + Vitamin K + Ca^{++} → [Prothrombin] +

Ca^{++} → [Thrombin] + [Fibrinogen]
⋮
[Fibrin Threads]
+
R.B.C., W.B.C., and Platelets

Fibrin-stabilizing Factor → [CLOT]

The Blood, Lymphatic, and Cardiovascular Systems

27. Hemoglobin is produced in immature blood cells known as erythroblasts and normoblasts, which are found in bone marrow. Iron copper, μ-ketoglutaric acid, glycine and globin are necessary for synthesis of hemoglobin. Women normally have lower hemoglobin values due to a lower metabolic rate and less muscle mass than males.

28. a) **Prothrombin:** The inactive precursor to thrombin. Prothrombin is a protein formed in the liver and is transported through the circulatory system.

    b) **Fibrinogen:** An inactive protein produced continuously by the liver and converted into fibrin

    c) **Thromboplastin:** A protein that is released from injured or damaged tissues. Thromboplastin begins the extrinsic clotting reaction.

29. Rectum → Superior Hemorrhoidal Vein → Inferior Mesenteric Vein → Inferior Vena Cava → Right Atrium → Tricuspid Valve → Right Ventricle → Pulmonary Semilunar Valve → Pulmonary Artery → Lungs → Pulmonary Veins → Left Atrium → Bicuspid (Mitral) Valve → Left Ventricle → Aortic Semilunar Valve → Ascending Aorta → Aortic Arch → Descending Aorta → Inferior Mesenteric Artery → Superior Hemorrhoidal Artery → Rectum

30. a) left atrium
    b) bicuspid (mitral) valve
    c) myocardium
    d) endocardium
    e) pericardium
    f) coronary arteries
    g) aortic semilunar valve
    h) left ventricle
    i) septum

31. a) plasma
    b) monocytes
    c) erythrocytes
    d) platelets
    e) antibodies, lymphocytes
    f) anemia
    g) leukocytosis

ANSWERS TO CASE STUDIES

1. She incorrectly believes that her unborn child may suffer a problem because of what she was told about herself.

   As a neonate, Rosa was cyanotic because of lack of O_2 in her blood.

   No

   Rosa is Rh+ because she has given birth to a normal child. Whether her unborn child is Rh+ or Rh-, there will be no incompatibility because she is Rh+.

Unit 9

2. Tonsils are lymphoid tissue important in the body's immune defenses. They act as sites where many lymphocytes can congregate and become exposed to various antigens which interact with foreign substances.

3. Stan's symptoms are best explained by a neurological dysfunction secondary to his valve replacement.
   Because of his high blood cholesterol and his mechanical cardiac valve, he is at an increased risk to form emboli in the heart chambers or plaques on the valve itself. These emboli could move via the internal carotid artery to the brain, causing a stroke.

UNIT 10

The Respiratory System

ANSWERS TO QUESTIONS WITHIN EXERCISES

UNIT 10A - RESPIRATORY SYSTEM ANATOMY

10A.1. It traps dust particles before they enter the lungs.

10A.2. It provides flexibility.

10A.3. Lung: variable responses
Trachea or bronchus: variable responses

10A.4. They provide for flexibility in the event of injury.

10A.5. Three

10A.6. Two

10A.7. Diaphragm, external intercostals, scalenes, sternocleidomastoids, and pectoralis

10A.8. Diaphragm, external intercostals, scalenes, sternocleidomastoids, and pectoralis

10A.9. Quiet: occurs without effort
Labored: is forced and under more conscious control

10A.10. Cat: there are three lobes on the right, three on the left, and a caudate lobe posterior to the heart

10A.11. 1) Left carotid artery
2) Pulmonary artery
3) Left medial lobe
4) Right medial lobe
5) Right posterior lobe
6) Brachiocephalic artery

UNIT 10B - RESPIRATORY MEASUREMENT

10B.1. Variable responses (usually 3000 - 5500 cc)

10B.2. Variable responses

10B.3. Student/standard x 100 = % vital capacity

10B.4. Variable responses (usually 500 - 800 cc)

10B.5. Variable responses

10B.6. Variable responses

10B.7. Variable responses

10B.8. 1200 cc

10B.9. Variable responses (respiratory rate x TV)

10B.10. Variable responses

10B.11. Longer; there is more oxygen in the lungs and tissues

28 Unit 10

ANSWERS TO SELF-TEST QUESTIONS

(1)	c	(2)	b	(3)	b	(4)	d	(5)	c	(6)	b	(7)	d
(8)	d	(9)	c	(10)	b	(11)	c	(12)	a	(13)	c	(14)	b
(15)	b	(16)	c	(17)	d	(18)	d	(19)	a	(20)	c	(21)	b
(22)	b	(23)	a	(24)	a	(25)	a						

DISCUSSION QUESTIONS

26. Student's results

27. In the human being, the right lung is divided into three lobes, the left into two lobes. In the cat, the right lung has three lobes and the left has three lobes.

28. The major advantage of the epiglottis being cartilaginous is for protection during breathing and swallowing. As the larynx moves upward and forward during swallowing, the free edge of the epiglottis is moved downward, helping to close the opening of the larynx.

29. Simple squamous epithelium allows for diffusion of respiratory gases to and from the pulmonary capillaries and the alveolar sacs.

ANSWERS TO CASE STUDY

Jack's VC of 7000 ml is above average. The standard for his age and height is approximately 5000 ml of air. The exercise regimen involved in competitive swimming has probably been the major factor involved in increasing his lung capacity. Jack's TV (tidal volume) is 787.5 ml of air.

$$\frac{7000}{80} :: \frac{X(TLC)}{1000} = 8750 \text{ ml} = \text{total lung capacity (TLC)}$$

TV is 9% of TLC, or 8750 x .09 = 787.5
Standard TV (tidal volume) is approximately

$$\frac{5000 \text{ ml}}{80} :: \frac{X(TLC)}{1000} = 6250.0 = \text{TLC} \times .09 = 562.5 \text{ ml}$$

Many adults who have suffered childhood asthma have increased respiratory volumes because of compensatory mechanisms involved in breathing. Increased musculature and deep breathing are among them.

UNIT 11

The Digestive System

ANSWERS TO QUESTIONS WITHIN EXERCISES

UNIT 11A - DIGESTIVE ANATOMY

11A.1. Cardiac and pyloric

11A.2. Rugae - allow for distension and increased surface area
Villi - absorption of digested food

11A.3. Common bile duct and main pancreatic duct

11A.4. Salivary amylase

11A.5. Mucosa, submucosa, muscularis externa, serosa

11A.6. Goblet cells: secrete mucus
Simple columnar epithelium: absorbs water and nutrients and secretes enzymes

11A.7. Cecum, ascending colon, transverse colon, descending colon, sigmoid colon, rectum

11A.8. A vestigial organ; contains lymphocytes that are part of the immune system for combatting infection

11A.9. Branch of hepatic portal vein, branch of hepatic artery, bile duct

UNIT 11B - DIGESTIVE CHEMISTRY

11B.1. Biuret reaction

11B.2. Seliwanoff's test

11B.3. To change the pH to match that of the trypsin

11B.4. It is insoluble in water.

RESULTS OF PROTEIN, CARBOHYDRATE, AND FAT EXERCISES		
TEST	RESULT	CONCLUSION
Proteins:		
Biuret reaction	violet color obtained	albumin contains peptide bonds
Millon reaction	brick red color obtained	albumin contains tyrosine
Ninhydrin reaction	purple-red color obtained	glycine & alanine (or albumin) are (or contain) alpha amino acids
Heat coagulation of protein	egg white is coagulated	egg white is a protein
Digestion of protein	cooked egg white is digested	trypsin (or pancreatin) are enzymes which digest protein

\multicolumn{3}{	c	}{RESULTS OF PROTEIN, CARBOHYDRATE, AND FAT EXERCISES—*continued*}
TEST	RESULT	CONCLUSION
Carbohydrates:		
Starch test	color turns blue/purple	starch is present
Starch hydrolysis test	no color change	starch was broken down into glucose
Molisch reaction	pink ring formed at boundary of glucose solution + H_2SO_4	glucose is a carbohydrate
Benedict's test	green, red or yellow precipitate formed (not for sucrose)	glucose, fructose, mannose, maltose, and lactose are reducing sugars; sucrose is not a reducing sugar
Barfoed's test	red precipitate formed	glucose is a monosaccharide
Seliwanoff's test	fructose turns color, glucose does not	differentiates between glucose and fructose
Inversion of sucrose	nothing happens to first tube; second tube turns color	sucrose not a reducing sugar; adding acid and base to sucrose breaks disaccharide bond, resulting in fructose and glucose, which are reducing sugars
Fats:		
Solubility of fat	oil floats on water, mixes on agitation with others	oil not soluble in water, slightly soluble in ethyl alcohol, soluble in others
Digestion of emulsified fat	with trypsin, color changes from blue to white or pink; with $NaHCO_3$, color remains blue	trypsin digests fat to fatty acids, which changes pH to acidic, and blue litmus loses color; $NaHCO_3$ does not change pH to acidic

ANSWERS TO SELF-TEST QUESTIONS

(1) d   (2) d   (3) d   (4) d   (5) d   (6) d   (7) c
(8) c   (9) d   (10) d   (11) c   (12) a   (13) c   (14) b
(15) a   (16) d   (17) a   (18) c   (19) a   (20) a   (21) a
(22) c   (23) b   (24) d   (25) c

DISCUSSION QUESTIONS

26. To increase the amount of absorption in small intestine
27. (a) appendix - present in human, absent in cat
    (b) cecum - present in human and cat

(c) hepatic flexure - present in human and cat

(d) sigmoid colon - present in human, absent in cat

28. (a) amylase - salivary glands, pancreas

(b) pepsin - gastric glands

(c) trypsin - pancreas

(d) peptidase - pancreas, intestinal glands

(e) lactase - intestinal glands

(f) lipase - gastric glands, pancreas

ANSWERS TO CASE STUDIES

1. Michelle probably has some involvement of her gallbladder and/or biliary system. She could have stones in the gallbladder itself or the ducts.

    Her discomfort occurs after dinner because her meal most likely included lipids. Upon ingestion of these lipids (fats) her gallbladder empties bile into the small intestine. There, the bile should emulsify the lipids. The obstruction interferes with this function.

    Gallbladder, cystic duct, duodenum and perhaps the liver, hepatic, and common bile ducts.

2. The pancreas produces enzymes which are involved in the digestion of all three food types: proteins, carbohydrates and lipids. Trypsin digests proteases and peptones. Chymotrypsin and peptidases digest small polypeptides. Amylase digests starches. Lipase digests emulsified fats.

    George's pain is most likely due to pancreatic dysfunction. George's doctor is investigating his weight loss, which could be attributed to his pancreatic dysfunction.

    The pain is felt in the back because the pancreas lies in the lesser curvature of the stomach on the lateral surface of the duodenum, toward the dorsal peritoneum. His food is not being digested; therefore, he is fatigued and is reporting bowel changes as undigested food is being excreted.

3. The ingested milk remains in and distends the stomach, causing the projectile vomiting.

    The new mother was right to be concerned. Her young son probably has pyloric stenosis, in which the pyloric sphincter muscles are not fully developed. Correction can be made by surgical repair.

UNIT 12

The Urinary System

ANSWERS TO QUESTIONS WITHIN EXERCISES

UNIT 12A - URINARY SYSTEM ANATOMY

12A.1. No

12A.2. Renal corpuscle - glomerulus and Bowman's (glomerular) capsule
Renal capsule - dense fibrous connective tissue covering the kidney

12A.3. Collecting duct → papillae → calyx → renal pelvis → ureter → bladder → urethra → urethral orifice

12A.4. Calyces; pyramids

12A.5. Variable responses

UNIT 12B - URINARY SYSTEM PHYSIOLOGY

12B.1. Pyelonephritis, glomerulonephritis, stress, renal failure, multiple myeloma

12B.2. Diabetes mellitus, Cushing's syndrome, pheochromocytoma

12B.3. Cirrhosis; hepatitis

ANSWERS TO SELF-TEST QUESTIONS

(1) c	(2) a	(3) b	(4) b	(5) a	(6) c	(7) d
(8) a	(9) a	(10) b	(11) b	(12) d	(13) c	(14) d
(15) d	(16) c	(17) c	(18) a	(19) a	(20) b	(21) c
(22) b	(23) d	(24) b	(25) b			

DISCUSSION QUESTIONS

26. Bowman's capsule being one cell layer in thickness allows for filtration of materials from blood vessels (glomerulus) to the capsule.

27. The urethra of the male is urogenital in function. The female urethra is only urinary in function. The male urethra is also longer than the female's.

28. Turbidity is a measure of particles dissolved and/or suspended in the urine. Specific gravity is a comparative measure of the density of the urine. Therefore, the greater the turbidity, the greater the specific gravity.

29. Urinalysis is considered a valuable diagnostic test because various substances which are necessary for a normally functioning body are excreted into the urine in various illnesses in increased or decreased amounts.

30. The kidneys are supplied with visceral afferent nerve fibers which transmit impulses from the kidneys to the CNS. Vasoconstrictor fibers from the thoracolumbar division of the ANS innervate the renal vascular system, thus controlling the amount of blood circulating through the glomeruli and the amount of glomerular filtrate formed.

ADH (Antidiuretic hormone) - is involved with the regulation of water balance; is produced by the hypothalamus and secreted by the neurohypophysis, and acts on the cells of the collecting tubules to increase water absorption.

Aldosterone - produced by the adrenal cortex, influences active transport of sodium and potassium by retaining sodium and excreting potassium from the distal convoluted tubules. When aldosterone deficiency exists, potassium is retained by the body while sodium, with an equivalent amount of water, is excreted into the urine. Body fluids can be reduced to dangerously low levels.

Renin - produced by the juxtaglomerular cells of nephrons when a reduced renal blood supply exists. Renin, upon entering the blood, acts on angiotensinogen, converting it to angiotensin I, which is subsequently converted to angiotensin II. Angiotensin II acts to increase renal blood pressure and blood flow. It also increases peripheral resistance by causing generalized vasoconstriction of arterioles in other parts of the body and by stimulating aldosterone production.

ANSWERS TO CASE STUDY

Urinalysis reveals that David Donne has an elevated glucose and ketones. His urine is also slightly acid - pH 6. These combined lab outcomes indicate the need for further clinical testing to determine the possibility of diabetes mellitus.

Glucose is not being transported into the cells and is spilling into the urine as a higher than usual rate. The nephrons are unable to reabsorb all the excess.

The fatty acids are being metabolized in increased amounts because of the inability of the cells to utilize glucose; therefore, ketones in increased amounts are being formed as a fatty acid breakdown. Ketonic acid accounts for the lower pH and the acidity in both urine and blood.

UNIT 13

Acid-Base Balance

ANSWERS TO QUESTIONS WITHIN EXERCISES

UNIT 13A - THE MEASUREMENT OF pH

13A.1. Seven

13A.2. Saliva contains uric acid and bicarbonate ions; urine contains buffer systems that compensate for pH changes in blood.

13A.3. Yes, urea decomposes into ammonia upon standing.

UNIT 13B - THE REGULATION OF pH

13B.1. Buffers resist changes in pH.

13B.2. To minimize contamination of equipment; more precise readings can be obtained for distilled water and buffer solutions, where pH changes are more subtle

13B.3. Step 2) hyperventilation results in an excess of O_2 in the blood and CO_2 being blown off, increasing pH of blood; slowing respirations so CO_2 is allowed to re-equilibriate, lowering pH of blood to normal

Step 3) the paper bag traps CO_2 so the subject is rebreathing it instead of O_2, decreasing pH of the blood; when normal breathing resumes, more O_2 is inhaled, respirations are faster, and pH of the blood is increased to normal

13B.4. Variable responses

ANSWERS TO ACID-BASE BALANCE PROBLEMS

1. Compensated metabolic alkalosis
2. Compensated respiratory acidosis
3. Uncompensated metabolic acidosis
4. Uncompensated respiratory alkalosis
5. Uncompensated respiratory and metabolic acidosis

ANSWERS TO SELF-TEST QUESTIONS

(1) b   (2) d   (3) a   (4) c   (5) d   (6) b   (7) a
(8) c   (9) c   (10) a (11) c (12) b (13) d (14) a
(15) b (16) b (17) c (18) a (19) F (20) T (21) F
(22) T (23) T (24) F (25) F

Acid-Base Balance

DISCUSSION QUESTIONS

26. Ranking from acidic to basic:
    freshly voided urine
    plasma
    saliva
    stagnant urine

27. Properties of Acids:
    1. hydrogen ion donor
    2. in solution, have a pH less than 7
    3. turn blue litmus red

    Properties of Bases:
    1. hydrogen ion acceptor
    2. in solution, have a pH greater than 7
    3. turn red litmus blue

28. Kidney Regulation:

    Hydrogen ions are secreted into the kidney tubules and bicarbonate ions are retained in the blood in response to an acidic (low) pH, resulting in raising the pH of the blood.

    Bicarbonate ions are secreted into the kidney tubules and hydrogen ions are retained in the blood in response to an alkaline pH, resulting in lowering the pH of the blood.

    Respiratory Center Regulation:

    The respiratory center in the medulla increases respiratory rate in response to an increase in arterial blood pCO_2. Peripheral chemoreceptors in the carotid and aortic bodies increase respiratory rate in response to decreased arterial pO_2. These mechanisms raise the blood pH.

    Shallow breathing or decreased respiratory rate results in raising the arterial pCO_2 and the pH of the blood will decrease.

ANSWER TO CASE STUDY

Sam would probably have respiratory acidosis because of the inability to exhale CO_2. The CO_2 would be elevated because the elastic recoil of his alveoli is diminished, impeding the exhalation of CO_2. Over time and in a chronic condition, metabolic alkalosis compensation would occur. His kidneys would excrete hydrogen ions and reabsorb more bicarbonate ions.

UNIT 14

The Reproductive System

ANSWERS TO QUESTIONS WITHIN EXERCISES

UNIT 14A - REPRODUCTIVE SYSTEM ANATOMY

14A.1. The foreskin, or prepuce, of the penis

14A.2. They both contain two uterine horns; the serosa of the non-pregnant pig uterus is corrugated in appearance.

14A.3. The epididymis

14A.4. 46

14A.5. 23, duplicate

14A.6. Secondary oocyte

14A.7. Primary spermatocytes and primary oocytes

14A.8. The cat uterus is bicornate in order to facilitate litters (multiple gestations); the human uterus is unicornate (designed for single gestations).

14A.9. Seminiferous tubules → epididymis → vas deferens → ejaculatory duct → urethra

UNIT 14B - REPRODUCTIVE SYSTEM PHYSIOLOGY

14B.1. End of the eighth week of gestation

14B.2. Variable responses

14B.3. Variable responses

ANSWERS TO SELF-TEST QUESTIONS

(1) c   (2) d   (3) a   (4) a   (5) c   (6) b   (7) a
(8) a   (9) d   (10) c (11) c (12) d (13) c (14) b
(15) d (16) c (17) b (18) b (19) b (20) a (21) b
(22) d (23) c (24) b (25) b

DISCUSSION QUESTIONS

26. seminiferous tubules epididymis ductus deferens (or vas deferens) ejaculatory duct urethra

27. Graafian follicle (in ovary) Fallopian tube (or uterine tube) Uterus (where implantation occurs)

28. It becomes a corpus albicans if fertilization does not take place. If fertilization occurs, the corpus luteum continues to secrete hormones.

29. (a) haploid
    (b) diploid
    (c) haploid
    (d) diploid
    (e) haploid
30. FSH and LH

ANSWERS TO CASE STUDY

Regardless of implantation, Tara's ovaries are unable to produce viable ova or hormones. Tara's endometrium, as a result of hormone therapy, was receptive to embryo implantation and maintenance. Prior to hormonal therapy, Tara's FSH and LH levels were high. After hormone therapy, they were within normal levels. FSH and LH levels are controlled through negative feedback by the synthetically replaced progesterone and estrogen.

UNIT 15

The Endocrine System

ANSWERS TO QUESTIONS WITHIN EXERCISES

UNIT 15A - THE ENDOCRINE GLANDS

15A.1. It contains both exocrine and endocrine tissue.

15A.2. Prolactin: initiates and maintains milk secretion (production) by mammary glands
Oxytocin: stimulates ejection of milk from mammary glands; also stimulates uterine contractions during labor

15A.3. ADH - stimulates water reabsorption in collecting ducts of kidneys; constricts arterioles
Oxytocin - stimulates uterine contractions during labor and stimulates lactation

15A.4. Decreases blood levels of calcium of opposing PTH action

15A.5. Principal cells - secrete PTH

15A.6. Cortisol; aldosterone

UNIT 15B - ENDOCRINE SYSTEM PHYSIOLOGY

15B.1. Variable responses

15B.2. Variable responses

15B.3. mg/dL or mg%

15B.4. Glucagon - elevates blood sugar by converting glycogen into glucose

ANSWERS TO SELF-TEST QUESTIONS

(1) d	(2) a	(3) b	(4) b	(5) a	(6) e	(7) e
(8) c	(9) b	(10) a	(11) b	(12) d	(13) c	(14) c
(15) b	(16) b	(17) a	(18) b	(19) c	(20) c	(21) b
(22) b	(23) d	(24) F	(25) T			

DISCUSSION QUESTIONS

26. (a) tetany
   (b) acromegaly
   (c) diabetes mellitus
   (d) Cushing's disease
   (e) Addison's disease
   (f) hypothyroidism (or myxedema)
   (g) goiter
   (h) cretinism

27. No, it becomes part of the neurohypophysis.
28. Thyroglobulin is the storage form of thyroxine.
29. No, colloid stores would become depleted and the epithelium would be columnar.

ANSWER TO CASE STUDY

Mary's basic problem is that, without insulin, glucose cannot cross the cell membranes, especially cell membranes of skeletal muscle and adipose tissue. Insulin is a protein that acts as a carrier molecule of glucose transport across cell membranes. Because glucose draws water out of cells, Mary's urine volume increases. Her cells become dehydrated, and she will become thirsty.

List of Materials and Supplies

UNIT 1

Medical Terminology

MATERIALS NEEDED PER CLASS OF 32 STUDENTS

EXERCISE 1   TERMS DESCRIBING HUMAN BODY DIRECTIONS

MATERIALS

1 or 2 Human torsos

Human anatomical charts

EXERCISE 2   TERMS DESIGNATING PLANES OF THE BODY

MATERIALS

1 or 2 Human torsos

Human anatomical charts

EXERCISE 3   TERMS DESIGNATING MAJOR BODY CAVITIES AND REGIONS

MATERIALS

1 or 2 Human torsos

Human anatomical charts

EXERCISES 4 THROUGH 9

MATERIALS

None

UNIT 2

The Microscope

MATERIALS NEEDED PER CLASS OF 32 STUDENTS

EXERCISE 1   THE STRUCTURE OF THE COMPOUND MICROSCOPE

MATERIALS (for a class working in pairs)

Compound microscopes

Lens paper

EXERCISE 2   FUNCTIONS OF THE MICROSCOPE PARTS

MATERIALS

Same as Exercise 1

EXERCISE 3   PROPER CARE OF THE MICROSCOPE

MATERIALS

Same as Exercise 2

EXERCISE 4   MICROSCOPIC OBSERVATION OF PREPARED SLIDES

MATERIALS

Compound microscopes

Prepared slides of letter "e" and other specimens (thyroid gland, simple columnar epithelium, skeletal muscle suggested)

EXERCISE 5   MICROSCOPIC OBSERVATION OF A HUMAN HAIR

MATERIALS

Compound microscopes

Roll of Scotch Brand Magic Tape

Strand of hair (from students)

UNIT 3

Cells

MATERIALS NEEDED PER CLASS OF 32 STUDENTS

A. **CELL STRUCTURE**

    EXERCISE 1 ANIMAL CELL STRUCTURE

        MATERIALS

        1 or 2 Animal cell models

    EXERCISE 2 OBSERVATION OF ORAL MUCOSA CELLS

        MATERIALS (for a class working in pairs)

        16 Compound microscopes

        32 Pre-cleaned microscope slides

        32 Cover slips

        32 Flat toothpicks

        16 Dropper bottles of methylene blue

        16 Wash bottles containing deionized or distilled water

        Sealed, puncture-proof container

    EXERCISE 3 MITOSIS

        MATERIALS (for a class working in pairs)

        16 Compound microscopes

        16 Prepared slides of whitefish mitosis

B. **CELL PHYSIOLOGY**

    EXERCISE 1 DIFFUSION

        MATERIALS (for demonstration & a class working in pairs)

        1 Bottle of perfume

        1800 ml deionized or distilled water in 2L Erlenmeyer flask

        16 Potassium permanganate crystals ($KMnO_4$)

        16 Test tubes containing gelatin

16 Dropper bottles of methylene blue

16 Sharp probes

8 Test tube racks

16 (600 mL) beakers

EXERCISE 2　OSMOSIS

MATERIALS (for a class divided into 2 groups)
(Set-up at beginning of class period)

2 Large white potatoes

1 Drill bit (½ to ¾ inch in diameter)

1 Bottle molasses or Karo syrup

2 (1 meter) lengths glass tubing (capillary tubing is preferable)

2 Rubber stoppers, size 4

2 Meter sticks

2 (1000 mL) Beakers

2 Ring stands

2 Single adjustable Buret clamps

2 Paper towels

Water

EXERCISE 3　DIALYSIS

MATERIALS (for a class divided into 4 groups)

4 (100 mL) Beakers containing sodium chloride 5%:
　　(5 g NaCl/100 mL H_2O)

4 (100 mL) Beakers containing glucose 10%:
　　(10 g glucose/100 mL H_2O)

4 (100 mL) Beakers containing albumin
　　(heat 0.25 tsp. albumin/100 mL H_2O; or 1 egg white)

4 (8-inch lengths) dialysis tubing soaked in cold H_2O several
　　hours prior to class use

4 (10-inch lengths approximately) Glass tubing

8 (600 mL) Beakers; 4 containing distilled water;
　　4-6 Beakers for water baths

4 Small dropper bottles of dilute nitric acid (HNO_3)

4 Small dropper bottles of silver nitrate (AgNO$_3$)

4 Small (30 mL) dropper bottles of Benedict's reagent

6-inch lengths of string or thread

24 Test tubes

4 Test tube racks

12 Graduated cylinders (50 mL)

4 Ring stands

4 Bunsen burners or hot plates

12 each (1 mL & 5 mL) Pipettes

4 Asbestos pads for Bunsen burners

Flint Strikers (if using Bunsen burners)

Safety glasses

Suggested CD-ROM The *Virtual Physiology Lab* #9 Diffusion, Osmosis & Tonicity.

EXERCISE 4 HYPERTONIC, ISOTONIC, AND HYPOTONIC SOLUTIONS

MATERIALS (for a class working in groups of 4)

24 Pre-cleaned microscope slides

Sheep red blood cells (optional)

Alcohol wipes

Sterile lancets (if sheep red blood cells not available)

24 Cover slips

8 Dropper bottles of 1.5% saline solution (1.5 g NaCl/100 cc deionized or distilled water)

8 Dropper bottles of distilled water

8 Dropper bottles of 0.9% normal saline (0.9 g NaCl/100 cc deionized or distilled water)

8 Wax pencils

8 Compound microscopes

Flat toothpicks

Paper towels

Disposable, non-sterile gloves

Disinfectant solution

Sealed, puncture-proof container

EXERCISE 5 ENZYMES

MATERIALS

CD-ROM *Virtual Physiology Lab* #10 Enzyme Characteristics

UNIT 4

Tissues

MATERIALS NEEDED PER CLASS OF 32 STUDENTS

EXERCISE 1 MICROSCOPIC IDENTIFICATION OF TISSUE TYPES

MATERIALS (for a class working in groups of 2 to 4)

16 Compound microscopes

8 Microscope slide boxes containing the following prepared tissue types:

- Simple squamous epithelium
- Simple cuboidal epithelium
- Simple columnar epithelium
- Stratified squamous epithelium
- Pseudostratified ciliated columnar epithelium
- Transitional epithelium
- Dense, fibrous connective tissue
- Areolar tissue
- Adipose tissue
- Reticular tissue
- Fibrocartilage
- Hyaline cartilage
- Elastic cartilage
- Osseous tissue (Compact bone)
- Skeletal muscle
- Smooth muscle
- Cardiac muscle
- Cardiac muscle, stained to show intercalated disks

EXERCISE 2 MICROSCOPIC IDENTIFICATION OF THE SKIN

MATERIALS (for a class working in groups of 2 to 4)

16 Compound microscopes

8 Microscope slide boxes containing the following prepared slides:

Skin - White

Skin - Pigmented

Skin, showing hair follicles (scalp)

Skin, showing Pacinian corpuscles

Skin, showing sebaceous and sweat glands

EXERCISE 3 HUMAN SKIN

MATERIALS (for a class working in 4 groups)

Models of human skin

EXERCISE 4 HUMAN FINGERPRINTS

MATERIALS (for a class working in groups of 4)

16 No. 2 pencils

32 Sheets of white paper (copier paper)

4 Rolls ¾" Scotch Brand Magic Tape

8 Dissecting microscopes or magnifying glasses

UNIT 5

The Skeletal System

MATERIALS NEEDED PER CLASS OF 32 STUDENTS

A. SKELETAL ANATOMY

    EXERCISE 1   MICROSCOPIC EXAMINATION OF OSSEOUS TISSUE

        MATERIALS (for a class working in pairs)

        16 Compound microscopes

        16 Prepared slides of dry ground bone (osseous tissue)

    EXERCISE 2   GROSS ANATOMY OF A LONG BONE

        MATERIALS (demonstration)

        1-2 Sagittally cut long bones

    EXERCISE 3   IDENTIFICATION OF BONES OF THE HUMAN SKELETON

        MATERIALS (for a class working in groups of 4)

        8 Disarticulated human skeleton sets

        1 Labeled adult skull

        8 Adult human skulls

        1-2 Articulated human skeletons

        1-2 Preparations of ear ossicles (embedded in plexiglass)

    EXERCISE 4   IDENTIFICATION OF BONE MARKINGS

        MATERIALS (for a class working in groups of 4)

        8 Disarticulated human skeleton sets

        8 Adult human skulls

        1 Labeled adult skull

        1-2 Beauchene preparation skull models

        8 Disarticulated human vertebrae sets

        4 Articulated human vertebral columns

        4 Articulated human vertebral columns with pelvis

        Human anatomical charts

B. **IDENTIFICATION OF JOINTS**

    EXERCISE 1   SYNARTHROSES

        MATERIALS (for a class working in 4 groups)
        MODELS OF EACH OF THE FOLLOWING:

        4 Human adult skulls

        1-2 Human fetal skulls

        1-2 Articulated skeletons

        4 Articulated human vertebral columns

    EXERCISE 2   DIARTHROSES

        MATERIALS (for a class working in 4 groups)

        Models of hip, knee and elbow joints

        4 Articulated skeletons

UNIT 6

The Muscular System

MATERIALS NEEDED PER CLASS OF 32 STUDENTS

A. **HUMAN MUSCULATURE**

   EXERCISE 1   MICROSCOPIC IDENTIFICATION OF MUSCLE TYPES AND MYONEURAL JUNCTIONS

   MATERIALS (for a class working in groups of 4)

   16 Compound microscopes

   8 microscope slide boxes containing the following prepared slides:

   Skeletal muscle

   Smooth muscle

   Cardiac muscle (preferably stained to show intercalated disks)

   Myoneural junction

   EXERCISE 2   IDENTIFICATION OF MUSCLES

   MATERIALS (Demonstration/class working in 4 groups)

   Anatomical charts of human musculature

   4 Human torso models

   4 Human eye models

   Disposable gloves

   4 Human head, arm and leg models

   Prosected cadaver (if available)

B. **CAT MUSCULATURE**

   EXERCISE 1   SKINNING THE CAT

   MATERIALS (for a class working in pairs)

   16 Preserved cats

   Disinfectant solution

   Dissecting instruments

   Dissecting pins

   16 Dissecting trays

Disposable gloves

Newspapers

Paper towels

Large plastic storage bags

Twist ties

Name tags with fastening wires

Preservative solution [140 mL BIO-PERM (available from Sargent-Welch): 860 mL water]

Safety glasses

EXERCISE 2 IDENTIFICATION OF MUSCLES

MATERIALS (for a class working in pairs)

16 Preserved cats

Dissecting instruments

Dissecting trays

Newspapers

Glycerol-Lysol preservative solution

Paper towels

Large plastic storage bags

Twist ties

Name tags with fastening wires

Disposable gloves

8 Dropper bottles of methyl salicylate (optional)

Cotton-tipped applicators (optional)

Safety glasses

Disinfectant solution

Large sponges

C. THE PHYSIOLOGY OF MUSCLE CONTRACTION

EXERCISE 1 CHEMISTRY OF MUSCLE CONTRACTION

MATERIALS (for a class working in pairs)

16 Pre-cleaned microscope slides

8 Psoas muscle preparation kits

1 Large (6-inch) Petri dish on ice (in which to separate muscle fibers)

Unit 6

ATP, KCl & MgCl$_2$ solutions from kit

16 Disposable Pasteur pipettes with bulbs

16 millimeter rulers

Containers for used pipettes

4 Small bottles of glycerol

Glass probes or straight probes (wooden-handled needles)

EXERCISE 2 INDUCTION OF FROG - SKELETAL MUSCLE CONTRACTION

MATERIALS

CD-ROM *Virtual Physiology Lab* #3 Frog Muscle

UNIT 7

The Nervous System

MATERIALS NEEDED PER CLASS OF 32 STUDENTS

A. NERVOUS SYSTEM ANATOMY

    EXERCISE 1   CYTOLOGICAL CHARACTERISTICS OF NEURONS

          MATERIALS (for a class working in groups of 2 to 4)

          16 Compound microscopes

          8 Prepared slides of motor neurons (spinal cord nerve smears)

          Neuron models

    EXERCISE 2   MICROSCOPIC EXAMINATION OF CEREBRUM AND CEREBELLUM

          MATERIALS (for a class working in groups of 2 to 4)

          16 Compound microscopes

          8 Prepared slides of cerebrum

          8 Prepared slides of cerebellum

    EXERCISE 3   MICROSCOPIC EXAMINATION OF THE SPINAL CORD

          MATERIALS (for a class working in groups of 2 to 4)

          16 Compound microscopes

          8 Prepared slides of cross section of spinal cord

          Spinal cord models

          2 Dissecting microscopes

    EXERCISE 4   EXAMINATION OF MENINGES

          MATERIALS (for a class working in groups of 4)

          8 Preserved segments of ox spinal cords

          8 Preserved sheep or beef brains with meninges

          8 Dissecting pans

          Dissecting instruments

          Disposable gloves

Unit 7

EXERCISE 5 EXAMINATION OF A BRAIN

MATERIALS (for a class working in groups of 4)

8 Preserved sheep or beef brains

8 Human brain models

8 Dissecting pans

Dissecting instruments

Disposable gloves

EXERCISE 6 EXAMINATION OF CRANIAL NERVES

MATERIALS (for a class working in groups of 4)

8 Dissecting pans

8 Preserved sheep or beef brains

8 Human brain models

Disposable gloves

Dissecting instruments

EXERCISE 7 EXAMINATION OF SPINAL NERVES

MATERIALS (for class working in groups of 4)

8 Dissecting pans

8 Preserved segments of ox spinal cords

8 Preserved cats

Dissecting instruments

Cadaver, if available

B. **NERVOUS SYSTEM PHYSIOLOGY**

EXERCISE 1 HUMAN REFLEXES

MATERIALS (for a class working in groups of 4)

8 Penlights or flashlights

8 Rubber reflex hammers

32 Paper cups

Pencils

Water

Facial tissues or sterile cotton balls

Chair

EXERCISE 2 TESTS FOR HUMAN CRANIAL NERVE FUNCTION

MATERIALS (for a class working in groups of 4)

8 Tuning forks (512 cycles per second preferable)

Small individual screw top jars containing:

> Tobacco
> Ground coffee
> Cinnamon
> Pepper

} freshly prepared each term

Water

Facial tissues

60 Paper cups

4 Stopwatches

32 Cotton balls

Ice water

8 (50 mL) Bottles of 10% NaCl solution (10 g NaCl/100 mL water)

8 (50 mL) Bottles of 10% sugar solution (10 g sucrose/100 mL water)

100 Cotton-tipped applicators (for entire class)

32 Tongue depressors

Printed matter

Chair

8 Penlights or flashlights

EXERCISE 3 TESTS FOR HUMAN CEREBELLAR FUNCTIONS

MATERIALS

None

EXERCISE 4 NERVE IMPULSE TRANSMISSION

MATERIALS

CD-ROM *Virtual Physiology Lab* #1 Action Potential

CD-ROM *Virtual Physiology Lab* #2 Synaptic Transmission

UNIT 8

The Special Senses

MATERIALS NEEDED PER CLASS OF 32 STUDENTS

A. THE EYE AND VISION

    EXERCISE 1   DISSECTION OF THE EYEBALL

MATERIALS (for a class working in groups of 4)

8 Eye models

Dissecting instruments

8 Dissecting trays

8 Preserved sheep or cow eyes

Disposable gloves

    EXERCISE 2   MICROSCOPIC EXAMINATION OF THE EYE

MATERIALS (for a class working in groups of 4)

8 Compound microscopes

8 Microscope slides of sagittal sections of eyeball

4 Dissecting microscopes

    EXERCISE 3   VISUAL ACUITY

MATERIALS (for a class working in groups of 4)

8 Snellen charts (or Figure 8.7 in lab manual)

    EXERCISE 4   LOCATION OF THE BLIND SPOT

MATERIALS

None (see lab manual p. 248)

    EXERCISE 5   TESTS FOR COLOR BLINDNESS

MATERIALS

Color-blindness charts (Ishihara plates recommended)

The Special Senses

B. **THE EAR, HEARING, AND EQUILIBRIUM**

   EXERCISE 1   ANATOMY OF THE EAR

   MATERIALS (for a class working in groups of 4)

   8 Ear models

   Anatomical charts of the human ear

   EXERCISE 2   MICROSCOPIC EXAMINATION OF THE EAR

   MATERIALS (for a class working in groups of 4)

   8 Compound microscopes

   4 Dissecting microscopes (optional)

   Microscope slides of middle and inner ear

   EXERCISE 3   HEARING TESTS

   MATERIALS (for a class working in groups of 4)

   4 Mechanical stopwatches

   8 Yardsticks/meter sticks

   Masking tape

   8 Rubber reflex hammers

   8 Tuning forks (512 cycles per second preferred)

   Chair

   EXERCISE 4   NYSTAGMUS

   MATERIALS (for a class working in groups of 3 or 4)

   1-2 Swivel armchairs

C. **OLFACTORY, TASTE, AND CUTANEOUS RECEPTORS**

   EXERCISE 1   OLFACTORY DISCRIMINATION

   MATERIALS (for a class divided into 4 groups)

   4 Human skulls

   Midsagittal section of human head model

   Samples of spices, selected from: nutmeg, cinnamon, vanilla extract, mustard powder, ginger, garlic powder, pepper, oregano, cloves, paprika, onion powder, and coffee

   Slices of apple and potato

Unit 8

EXERCISE 2 TASTE

MATERIALS (for a class working in groups of 4)

8 Compound microscopes

8 Prepared slides of tongue with taste buds

8 (25 mL) Dropper bottles of Menthol eucalyptus

225 Cotton-tipped applicators

40 PTC paper strips

10% Sugar solution (sucrose) (60 mL per class)

10% NaCl solution (60 mL per class)

Lemon juice

Mustard (not powdered)

60 Paper cups

20 Styrofoam cups

36 Ice cubes

Hot water (105 - 110°F)

EXERCISE 3 CUTANEOUS RECEPTORS

MATERIALS (for a class working in groups of 4)

8 Prepared slides of skin with sensory receptors

8 (50 mL) jars of powdered charcoal

Hot water (105 - 110°F)

8 Finishing nails or blunt probes (optional)

16 Large dissecting pins

Ice cubes (or crushed ice)

8 Centimeter rulers

8 Black or blue ballpoint pens

8 Compound microscopes

8 Small corks (or pieces of cork)

4 Skin models

8 Bunsen burners (optional)

UNIT 9

The Blood, Lymphatic, and Cardiovascular Systems

MATERIALS NEEDED PER CLASS OF 32 STUDENTS

A. ANATOMY OF THE BLOOD, LYMPHATIC SYSTEM, BLOOD VESSELS, AND HEART

    EXERCISE 1   MICROSCOPIC EXAMINATION OF BLOOD

MATERIALS (for a class working in groups of 2 to 4)

16 Compound microscopes

Blood cell models

8 microscope slide boxes containing the following prepared slides:

    Normal human blood smear (Wright's stain)

    Sickle cell anemia blood smear

    Leukemia blood smear

    Frog blood smear

4 Wash bottles containing distilled or deionized water

Synthetic mounting medium

Immersion oil

64 Rectangular cover slips

8 Wax pencils

64 Pre-cleaned microscope slides

32 Sterile lancets or hemolets

2 Automatic finger pricking devices (optional)

Alcohol wipes

Paper towels

8 Staining jars and staining racks

8 (50 cc) Wright's blood stain

8 (50 cc) Dropper bottles containing Wright's buffer solution

Disinfectant solution

Puncture-resistant, sealed disposal container

Disposable gloves

62 Unit 9

EXERCISE 2 THE LYMPHATIC SYSTEM

MATERIALS (for a class working in groups of 4)

8 Microscope slide boxes containing the following prepared slides:

    lymph node

    lymphatic vessel

    ileum portion of small intestine

    palatine tonsil

    spleen

    thymus gland

Models of human lymph node

Preserved cats

Demonstration cat with injected lymphatic system (optional)

Human cadaver (optional)

Dissecting kits

Dissecting pans

Newspaper

Disposable gloves

EXERCISE 3 MICROSCOPIC EXAMINATION OF AN ARTERY, VEIN, CAPILLARY, AND NERVE

MATERIALS (for a class working in pairs)

16 Compound microscopes

8 microscope slide boxes containing the following prepared slides:

    artery

    vein

    nerve

    capillary

Models of arteries, veins, and capillaries

EXERCISE 4 THE HUMAN HEART

MATERIALS (for a class working in groups of 4)

8 Human heart models

Human cadaver (optional)

The Blood, Lymphatic, and Cardiovascular Systems

EXERCISE 5 DISSECTION OF THE SHEEP HEART

MATERIALS (for a class working in groups of 4)

8 Preserved sheep hearts

Scalpel

Blunt probe

8 Dissecting pans

Dissecting instruments

Straws or wooden probes

Disposable gloves

EXERCISE 6 DISSECTION OF THE CAT HEART

MATERIALS (for a class working in pairs)

16 Dissecting pans

16 Double-injected, preserved cats

Triple-injected, preserved cat demonstration specimens

Glycerol-Lysol preservative solution

Wash bottles containing deionized or distilled water

Dissecting instruments

Disposable gloves

Large plastic bags

Paper towels

Name tags with fastening wires

Newspaper

EXERCISE 7 DISSECTION OF THE CAT CIRCULATORY SYSTEM WITH REFERENCE TO THE HUMAN

MATERIALS (for a class working in pairs)

16 Double-injected, preserved cats

Dissecting instruments

Dissecting pans

Disposable gloves

Newspapers

Human torso

Glycerol-Lysol preservative solution

Unit 9

Paper towels

16 Large plastic bags

16 Name tags with fastening wires

Cadaver (if available)

EXERCISE 8 HUMAN FETAL CIRCULATION

MATERIALS

1 - 2 Models of human fetal circulatory system

Human anatomical charts of fetal circulation

B. HUMAN BLOOD PHYSIOLOGY

EXERCISE 1 DETERMINATION OF ABO AND Rh BLOOD TYPES

MATERIALS (for a class working in pairs)

Alcohol wipes

Paper towels

32 Sterile lancets

Flat toothpicks

8 Wax marking pencils

4 Vials each: anti-A, anti-B, and anti-D (Rh) sera

1 Light warming box

64 Pre-cleaned microscope slides

64 Cover slips

16 Compound microscopes

Disposable gloves

Disinfectant solution

Sponges

Puncture-resistant, sealed disposal container

EXERCISE 2 HEMOGLOBIN ESTIMATION

MATERIALS (for a class working in pairs)

Tallquist scale and test paper

Sterile alcohol wipes

32 Sterile lancets (Hemolets)

Disposable gloves

The Blood, Lymphatic, and Cardiovascular Systems

Paper towels

Hemoglobinometer with batteries

Hemolysis applicator

Puncture-resistant, sealed disposal container

EXERCISE 3 PULSE DETERMINATION

MATERIALS (for a class working in groups of 4)

Cardio Tach Series 4600 pulse rate monitor or similar electronic monitor

Plastic or ribbon tape (optional)

EXERCISE 4 BLOOD PRESSURE DETERMINATION

MATERIALS

8 Stethoscopes

8 Mercury or aneroid sphygmomanometers

Electronic blood pressure apparatus, such as Sphygmostat, produced by Technical Resources, Inc., Waltham, Massachusetts

C. CARDIAC MUSCLE PHYSIOLOGY

EXERCISE 1 THE HUMAN ELECTROCARDIOGRAM

MATERIALS (demonstration)

Cardio Tach with ECG module

ECG cable with grabber connections

6 Reusable or 12 disposable skin electrodes

Suggested CD-ROM *Virtual Physiology Lab* #5 Electrocardiogram

EXERCISE 2 NORMAL CARDIAC CYCLE OF THE FROG

MATERIALS

Suggested CD-ROM *Virtual Physiology Lab* #4 Effects of Drugs on the Frog Heart

UNIT 10

The Respiratory System

MATERIALS NEEDED PER CLASS OF 32 STUDENTS

A. RESPIRATORY SYSTEM ANATOMY

   EXERCISE 1   EXAMINATION OF PREPARED SLIDES OF THE LUNG AND TRACHEA

   MATERIALS (for a class working in groups of 2 to 4)

   16 Compound microscopes

   8 Microscope slide boxes containing the following prepared slides:

   lung

   trachea

   EXERCISE 2   ANATOMY OF HUMAN RESPIRATORY ORGANS

   MATERIALS (for a class working in groups of 4)

   4 Models of human larynx

   4 Models of human lung and trachea

   2 Human torsos

   Human skulls

   1-2 Models of sagittal section of human head

   Cadaver (optional)

   EXERCISE 3   SHEEP PLUCK

   MATERIALS (for a class working in groups of 4)

   8 Sheep pluck

   8 Dissecting pans

   Dissecting instruments

   Disposable gloves

   EXERCISE 4   CAT RESPIRATORY ORGANS

   MATERIALS (for a class working in pairs)

   16 Preserved cats

   16 Dissecting pans

   Dissecting instruments

Dissecting pins

Newspapers

Disposable gloves

Glycerol-Lysol preservative solution

B. **RESPIRATORY MEASUREMENT**

EXERCISE 1 SPIROMETRIC MEASUREMENT AND CALCULATION OF STANDARD RESPIRATORY VOLUMES

MATERIALS (for a class working in groups of 4)

8 Spirometers

32 Disposable mouthpieces

32 Alcohol wipes

Suggested CD-ROM *Virtual Physiology Lab* #6 Pulmonary Function

EXERCISE 2 DETERMINATION OF RESPIRATORY VARIATIONS

MATERIALS (demonstration)

Pneumograph

Pneumograph tubing

Paper cups

Water

Suggested CD-ROM *Virtual Physiology Lab* #7 Respiration and Exercise

UNIT 11

The Digestive System

MATERIALS NEEDED PER CLASS OF 32 STUDENTS

A. DIGESTIVE ANATOMY

    EXERCISE 1   HUMAN DIGESTIVE ANATOMY

        MATERIALS (for a class working in groups of 2 to 4)

        1 or 2 Models of human teeth
        Models of digestive organs
        Human torso
        8 Human skulls

    EXERCISE 2   CAT DIGESTIVE ANATOMY

        MATERIALS (for a class working in groups of 2 to 4)

        16 Preserved cats
        Disposable gloves
        Dissecting instruments
        Newspapers
        Dissecting pans

    EXERCISE 3   MICROSCOPIC EXAMINATION OF DIGESTIVE TISSUE

        MATERIALS (for a class working in groups of 2 to 4)

        8 Microscope slide boxes containing the following prepared slides:
            developing tooth
            dried tooth
            tongue
            parotid gland
            esophagus
            stomach
            small intestine
            large intestine
            gallbladder
            appendix
            liver
            rectum
            anus (anal canal)
        8 (16) Compound microscopes

B. DIGESTIVE CHEMISTRY

EXERCISE 1   TESTS FOR PROPERTIES OF PROTEINS

MATERIALS (for a class working in groups of 4)

8 (25 mL) Bottles 1% albumin solution or egg white

8 (25 mL) Bottles of 10% NaOH

8 (25 mL) Bottles of 1% $CuSO_4$

8 (25 mL) Bottles of Millon's reagent

12 (60 mL) Beakers

Solution of 0.1% glycine or alanine (1 mL ethanol + 99 mL H_2O + 0.1 g alanine or glycine)

8 (25 mL) Bottles of 0.1% freshly prepared ninhydrin

Undiluted egg white from 4 eggs/32 students

200 mL water baths set at 37°C (or electric water baths)

72 Test tubes/32 students

8 (600 mL) Beakers

8 Ring stands

50 mL 5% trypsin solution

8 Bunsen burners or hot plates

8 Asbestos pads

16 Test tube holders

8 Flint strikers

8 Test tube racks

10 Medicine droppers

10 (1 mL) Pipettes with bulbs/32 students

40 (10 mL) Pipettes with bulbs/32 students

8 (10 mL) Graduated cylinders

500 mL Buffer solution (pH 8)

Safety glasses

Deionized water

EXERCISE 2   TESTS FOR PROPERTIES OF CARBOHYDRATES

MATERIALS (for a class working in groups of 4)

8 (25 mL) Dropper bottles of 5% alcoholic solution of alpha naphthol

50 mL 5% Fructose solution

50 mL Concentrated H_2SO_4

8 (25 mL) Dropper bottles of Benedict's solution

8 (25 mL) Dropper bottles of 5% glucose solution

8 (25 mL) Dropper bottles of 5% sucrose solution

8 (25 mL) Dropper bottles of Barfoed's reagent

8 (25 mL) Dropper bottles of 5% fructose solution

8 (25 mL) Dropper bottles of 10% glucose solution

8 (25 mL) Dropper bottles of Seliwanoff's reagent

8 (25 mL) Dropper bottles of 1 M HCL

8 (25 mL) Dropper bottles of 1 M NaOH (40 g/L)

20 (60 mL) Beakers

40°C Water bath (may be electric)

200 mL Thin starch paste

8 (25 mL) Dropper bottles of Lugol's iodine

8 (600 mL) Beakers for water bath

100 Test tubes/32 students

8 Bunsen burners or hot plates

8 Asbestos pads

Ring stands

Strikers (if using Bunsen burners)

8 Test tube racks

16 Test tube holders

Safety glasses

8 (25 mL) Graduated cylinders

65 (10 mL) Pipettes/32 students

Safety glasses

EXERCISE 3 TESTS FOR PROPERTIES OF FATS

MATERIALS (for a class working in groups of 4)

8 (25 mL) Dropper bottles of fresh cottonseed oil

100 mL Distilled water

8 (25 mL) Dropper bottles of ethyl alcohol

100 mL Can of ether

8 (25 mL) Dropper bottles of Bottle of benzene

8 (25 mL) Dropper bottles of carbon tetrachloride

100 mL Sweet cream (fresh)

25 mL 5% Trypsin

25 mL 0.5% $NaHCO_3$

56 Test tubes/32 students

8 Test tube racks

40°C Oven or electric water bath

8 (25 mL) Dropper bottles of blue litmus solution (1%)

8 (25 mL) Graduated cylinders

16 (10 mL) Pipettes

12 (60 mL) Beakers

Thermometer

Safety glasses

Suggested CD-ROM *Virtual Physiology Lab* #8 Digestion of Fats

UNIT 12

The Urinary System

MATERIALS NEEDED PER CLASS OF 32 STUDENTS

A. URINARY SYSTEM ANATOMY

   EXERCISE 1   MICROSCOPIC EXAMINATION OF RENAL TISSUE

   MATERIALS (for a class working in groups of 2 to 4)

   16 Compound microscopes

   8 Kidney slides

   EXERCISE 2   GROSS ANATOMY OF A KIDNEY

   MATERIALS (for a class working in groups of 4)

   8 Preserved kidneys (sheep or beef)

   8 Dissecting pans

   Dissecting instruments

   Disposable gloves

   Nephron models

   Kidney models

   EXERCISE 3   ANATOMICAL FEATURES OF THE HUMAN URINARY SYSTEM

   MATERIALS (for a class working in groups of 4)

   4 Human torsos

   4 Human urinary system models

   EXERCISE 4   ANATOMICAL FEATURES OF THE MALE AND FEMALE CAT URINARY SYSTEMS

   MATERIALS (for a class working in groups of 4)

   8 Preserved cats (4 male, 4 female preferable)

   8 Dissecting pans

   Dissecting instruments

   Disposable gloves

B. **URINARY SYSTEM PHYSIOLOGY**

    EXERCISE 1   FORMATION OF URINE

        MATERIALS

        None

    EXERCISE 2   ROUTINE URINALYSIS

        MATERIALS (for a class working groups of 2 to 4)

        32 Covered plastic or glass specimen containers

        16 Compound microscopes

        32 Pre-cleaned microscope slides

        32 Cover slips

        Disposable Pasteur pipettes with bulbs

        "Unknown" synthetic urine (see preparation to follow)

        One box of Multistix Reagent Strips for Urinalysis (Ames Div., Miles, Inc.) urine test strips

        40 (60 mL) Beakers

        PREPARATION OF UNKNOWN URINE

        Synthetic urine - stock solution:

        1000 mL Distilled water

        25 g Urea

        10 g NaCl

        8 Drops Ammonium hydroxide

        4 Drops Sulfuric acid

        Add 2 drops yellow food coloring and ¼ drop red food coloring

        UNKNOWN A

        200 mL Stock urine

        25 mL Glucose solution (1%)

        UNKNOWN B

        200 mL Stock urine

        ½ - 1 tsp Egg albumin

        UNKNOWN C

        200 mL Stock urine

        2 mL acetoacetic acid (lithium salt), few grains

        UNKNOWN D

        200 mL Stock urine

        6-8 drops of blood

        UNKNOWN E

        200 mL Stock urine

        2 mL Concentrated HCl (use under hood)

Unit 12

EXERCISE 3   MEASUREMENT OF URINE SPECIFIC GRAVITY USING A REFRACTOMETER

MATERIALS

Urine specific gravity refractometer (Atago Uricon-N recommended)

Medicine dropper

Urine sample

Distilled water

Soft damp cloth

Soft dry cloth

EXERCISE 4   URINE SCREENING TESTS

MATERIALS (for a class working in groups of 2 to 4)

8 (25 mL) Dropper bottles of 10% acetic acid

8 (25 mL) Bottles of Benedict's reagent

8 (25 mL) Dropper bottles of concentrated nitric acid

8 (400 mL) Beakers

32 Urine specimen cups

Student urine samples in covered plastic or glass containers

16 (25 mL) Graduated cylinders

32 (10 mL) Pipettes

32 Glass Petri dishes or watch glasses

Medicine droppers

8 each: Synthetic urine samples - containing protein, glucose, urea, blood (See Exercise 2 preparation)

1 Box clean cover slips

72 Gross pre-cleaned microscope slides

60 Test tubes

Test tube brushes (for clean-up)

1 Box of #2 filter paper

8 Hot plates

16 Compound microscopes

40 (60 mL) Beakers

UNIT 13

Acid-Base Balance

MATERIALS NEEDED PER CLASS OF 32 STUDENTS

A. THE MEASUREMENT OF pH

   EXERCISE 1   MEASURING THE pH OF COMMON SOLUTIONS

   MATERIALS *(for a class working in groups of 4)*

   2 Vials red litmus paper
   2 Vials blue litmus paper
   1-2 pH meters
   Wash bottles with distilled or deionized water
   600 mL pH 7 buffer solution
   2 Vials pH paper (or nitrazine paper)
   200 mL Milk
   200 mL Coffee
   200 mL Cola
   200 mL Lemon juice
   200 mL Baking soda solution
   200 mL Distilled water
   40 (60 mL) Beakers
   8 (400 mL) Beakers

   EXERCISE 2   DETERMINING THE pH OF BIOLOGICAL SOLUTIONS

   MATERIALS *(for a class working in groups of 4)*

   2 (30 mL) Saliva samples [can be collected from student(s)]
   8 (100 mL) Freshly voided urine samples
   8 (100 mL) Urine samples left to stand one hour
   50 mL Blood plasma (obtained by centrifuging sheep whole blood for 5 minutes at 1800 rpm) (optional)
   8 Rolls pH paper (or nitrazine paper)
   8 (60 mL) Beakers

B. THE REGULATION OF pH

EXERCISE 1 BUFFER SYSTEMS

MATERIALS (demonstration or class divided into 2 groups)

1-2 pH Meters

2-4 Beakers with 100 mL each of:

    distilled water

    pH 7 buffer solution

2 (25 mL) Dropper bottles of 0.05 M HCl

2 (25 mL) Dropper bottles of 0.05 M NaOH

Wax pencils

8 (60 mL) Beakers

4 (250 mL) Beakers

2 Wash bottles containing deionized water

EXERCISE 2 MAINTAINING ACID-BASE BALANCE IN THE BODY

MATERIALS (for a class working in pairs)

16 Small paper bags

1-2 pH Meters

1-2 Wash bottles of deionized water

16 (250 mL) Beakers

16 Straws

Clock or watch with second hand

EXERCISE 3 DISORDERS OF ACID-BASE BALANCE: ACIDOSIS AND ALKALOSIS

MATERIALS

None

UNIT 14

The Reproductive System

MATERIALS NEEDED PER CLASS OF 32 STUDENTS

A. REPRODUCTIVE SYSTEM ANATOMY

    EXERCISE 1   HUMAN REPRODUCTIVE ORGANS

        MATERIALS (for a class working in groups of 4)

        4 Models of mid-sagittal sections of male and female pelvis

        4 Human torsos

        4 Models of human female breast

        Human anatomical charts

        Male and female cadavers (optional)

    EXERCISE 2   CAT REPRODUCTIVE ORGANS

        MATERIALS (for a class working in pairs)

        16 Preserved cats (male and female)

        16 Dissecting pans

        Dissecting instruments

        Disposable gloves

        Newspapers

    EXERCISE 3   OBSERVATION OF PREGNANT PIG UTERUS AND BULL TESTICLE

        MATERIALS (demonstration)

        1-2 Preserved specimens of bovine and/or pig: ovary, testis, uterus with pig fetus in situ

        Dissecting instruments

        Disposable gloves

        1-2 Dissecting pans

    EXERCISE 4   MICROSCOPIC STUDY OF AN OVARY AND UTERUS

        MATERIALS (for a class working in groups of 2 to 4)

        16 Compound microscopes

        1 or 2 Dissecting microscopes

78 Unit 14

8 Microscope slide boxes containing the following prepared slides:

    uterus (resting phase)

    uterus (secretory phase)

    uterus (menstrual phase)

    ovary (showing Graafian follicles)

    ovary (showing corpus luteum)

EXERCISE 5 MICROSCOPIC STUDY OF A TESTIS AND PENIS

MATERIALS (for a class working in groups of 2 to 4)

16 Compound microscopes

1-2 Dissecting microscopes

8 Microscope slide boxes containing the following prepared slides:

Testis (rat testis showing spermatogenesis preferable)

Penis

Sperm smear

B. REPRODUCTIVE SYSTEM PHYSIOLOGY

EXERCISE 1 THE FEMALE SEXUAL CYCLE

MATERIALS

None

EXERCISE 2 URINARY hCG AS AN INDICATOR OF PREGNANCY

MATERIALS (for a class working in 4 groups)

3-4 Hybritech™ Tandem® Icon® II hCG (Urine) Pregnancy Test Kits[1], each containing:

    Test cylinders

    Antibody conjugate (Bottle A)

    Substrate reagent (Bottle B: to be stored in the dark—should be colorless when used)

    Wash concentrate (Bottle C)

    Transfer pipettes

---

[1] Tandem Icon II hCG (Urine) Pregnancy Test Kits, manufactured by Beckman-Coulter, Inc.; may be purchased from Fisher Scientific Company.

The Reproductive System

2 Negative and positive urine controls containing hCG at concentrations of 0 and 50 mIU/mL respectively (available from distributor)

4 Clean plastic or glass containers containing fresh urine samples

2 (500 mL) Plastic wash bottles

Distilled or deionized water (For each new kit, mix the contents of Bottle C—wash concentrate—with sufficient distilled or deionized water to make 500 mL of wash solution, then pour this solution into the 500 mL wash bottle.)

Clock or watch with second hand

EXERCISE 3 PRENATAL DEVELOPMENT

MATERIALS (demonstration)

1-2 Models or preserved specimens of human placenta

Models of stages of human embryonic and fetal development

UNIT 15

The Endocrine System

MATERIALS NEEDED PER CLASS OF 32 STUDENTS

A. THE ENDOCRINE GLANDS

    EXERCISE 1   ANATOMY OF ENDOCRINE GLANDS

        MATERIALS (for entire class)

        2-3 Human torsos

        Human anatomical charts

        Models of endocrine glands

    EXERCISE 2   MICROSCOPIC STUDY OF THE ENDOCRINE GLANDS

        MATERIALS (for a class working in groups of 2 to 4)

        16 Compound microscopes

        2-3 Dissecting microscopes

        8 Microscope slide boxes containing the following prepared slides:

            pituitary (hypophysis)

            thyroid

            parathyroid

            pancreas

            adrenal

B. ENDOCRINE SYSTEM PHYSIOLOGY

    EXERCISE 1   DETERMINATION OF BLOOD GLUCOSE LEVEL: VISUAL METHOD

        MATERIALS (for a class working in pairs)

        1 Vial containing Chemstrip bG®[2] Test Strips (50 or 100 test strips per vial)

        32 Sterile lancets (may be used with an automatic finger-pricking device)

        32 Dry cotton or rayon balls

        Soap and water

        32 Alcohol wipes

---

[2]Registered trademark of Boehringer-Mannheim Diagnostics, 9115 Hague Road, Indianapolis, IN 46250.

The Endocrine System

Disposable gloves

Clock or watch with second hand

Sealed, puncture-proof disposal container

EXERCISE 2 DETERMINATION OF BLOOD GLUCOSE LEVEL USING AN BLOOD GLUCOSE MONITOR

MATERIALS (for class working in groups of 2 to 4)

1-2 Accu-Chek® II Blood Glucose Monitors

1 Vial containing Chemstrip bG® (50 or 100 test strips per vial)

32 Sterile lancets (may be used with automatic finger pricking device)

32 Dry cotton or rayon balls

Soap and water

32 Alcohol wipes

Disposable gloves

Large garbage bag

Paper towels

Sealed, puncture-proof disposal container

EXERCISE 3 INSULIN SHOCK IN FISH

MATERIALS (for a class working in groups of 4)

2 Vials of Humulin regular insulin (U-100)

16 (500 mL) beakers

8 (200 mL) beakers

1200 mL 5% glucose solution

8 small minnows or large guppies (preferable) or goldfish

2-3 fish nets

1-2 (U-100) insulin syringes

Sealed, puncture-proof disposal container

Major Vendors

Major Vendors

Aldrich Chemical Company
1001 West Saint Paul Avenue
Milwaukee, WI 53233
1-414-273-3850
1-800-558-9160
FAX: 1-800-962-9591

Anatomical Chart Co.
8221 Kimball Avenue
Skokie, IL 60076-2956
1-847-679-4700
1-800-621-7500
Order FAX: 1-847-674-0211
Personal FAX: 1-847-679-9155

Carolina Biological Supply Company
2700 York Road
Burlington, NC 27215
1-910-584-0381
1-800-334-5551
FAX: 1-800-222-7112

Central Scientific Company
3300 Cenco Parkway
Franklin Park, IL 60131
1-847-451-0150
1-800-572-3626
FAX: 1-800-262-3626

Denoyer-Geppert Science Company
5225 Ravenswood Avenue
Chicago, IL 60640
1-773-561-9200
1-800-621-1014
FAX: 1-773-561-4160

Fisher Scientific Company
1600 West Glenlake Avenue
Itasca, IL 60143
1-630-773-3050
1-800-766-7000
FAX: 1-800-926-1166
Voice Mail: 1-800-955-6666

Fisher Scientific Company (E.M.D.)
485 South Frontage Road
Burr Ridge, IL 60521
1-630-655-4410
1-800-955-7999
FAX: 1-800-955-0740

Frey Scientific
905 Hickory Lane
Mansfield, OH 44901
1-419-589-1900
1-800-225-3739
FAX: 1-800-589-1522

Fryer Company Inc.
11177 East Main Street
Huntley, IL 60142-7147
1-847-669-2000
FAX: 1-847-669-2056

Health EDCO
P.O. Box 21207
Waco, TX 76710
1-817-776-6461
1-800-299-3366
FAX: 1-817-751-0221

J & H Berge
4111 South Clinton Avenue
South Plainfield, NJ 07080-0310
1-908-561-1234
FAX: 1-908-561-3002

Kilgore International Inc.
36 West Pearl Street
Coldwater, MI 49036
1-517-279-9000
1-800-892-9999
FAX: 1-517-278-2956

Lapine Scientific Company
13636 Western Avenue
Blue Island, IL 60406
1-708-388-4030
1-800-205-6303
FAX: 1-708-388-4084

Major Vendors

Midwest Culture Service
1924 North Seventh Street
Terre Haute, IN 47804
1-812-232-6785

Nasco Science
901 Janesville Avenue

Fort Atkinson, WI 53538
1-414-563-2446
1-800-558-9595
FAX: 1-414-563-8296

Nystrom
3333 Elston Avenue
Chicago, IL 60618-5898
1-800-621-8086
FAX: 1-773-463-0515

Sargent-Welch
911 Commerce Court
Buffalo Grove, IL 60089
1-847-459-6625
1-800-727-4368
FAX: 1-800-676-2540

Sigma Chemical Company
P.O. Box 14508
St. Louis, MO 63178
1-800-325-3010
FAX: 1-800-325-5052

Software House International
1626 Country Lakes Drive, #202
Naperville, IL 60563
1-630-548-1572
FAX: 1-630-548-1573

StanBio Laboratories
2930 East Houston Street
San Antonio, TX 78202
1-210-222-2108
1-800-531-5535
FAX: 1-210-227-6367

Triarch Incorporated
P.O. Box 98
Ripon, WI 54971
1-414-748-5125
1-800-848-0810
FAX: 1-414-748-3034

Tyron Global Company
4500 Empire Way, Ste. 4
Lansing, MI 48917
1-517-322-2800
FAX: 1-517-322-2809

VWR Scientific
P.O. Box 66929
O'Hare A.M.S.
Chicago, IL 60666
1-800-932-5000
FAX: 1-630-879-6718

Wards Natural Science
5100 West Henrietta Road
Rochester, NY 14692
1-716-359-2502
1-800-962-2660
FAX: 1-800-635-8439

Wilkens-Anderson Company
4525 West Division Street
Chicago, IL 60651
1-773-384-4433
1-800-847-2222
FAX: 1-773-384-6260

Unlabeled Illustrations

The following illustrations are duplicated from *A Laboratory Textbook of Anatomy & Physiology*, Seventh Edition. The labels have been removed, but the lead lines for the labels remain. They can be duplicated for testing purposes or converted into overhead transparancies by photocopying on clear acetate sheets.

Figure 1.7 Viscera in relation to anatomical abdominal regions

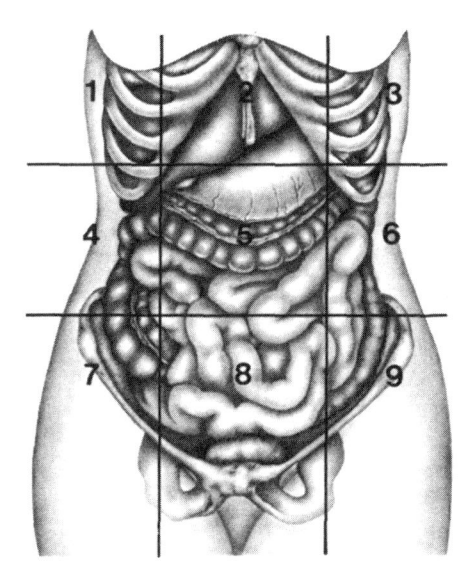

Figure 1.8 Viscera in relation to clinical abdominal regions

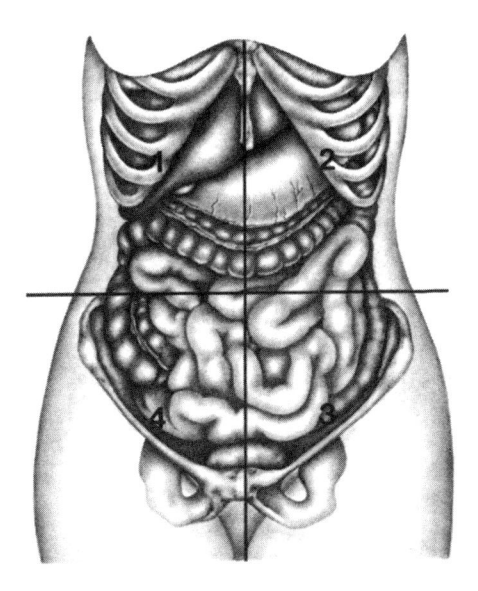

Figure 4.22 Thick skin, palm, showing epidermal layers, h.p.

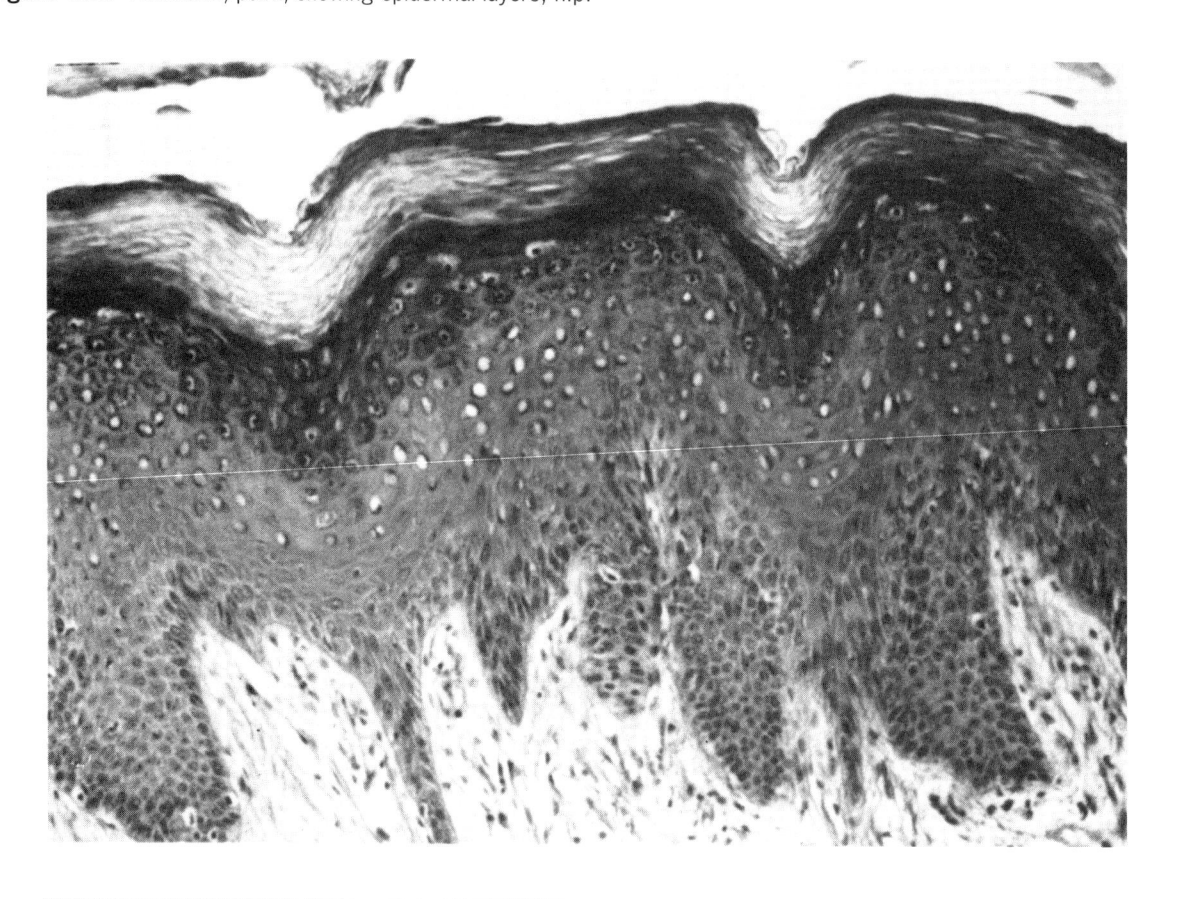

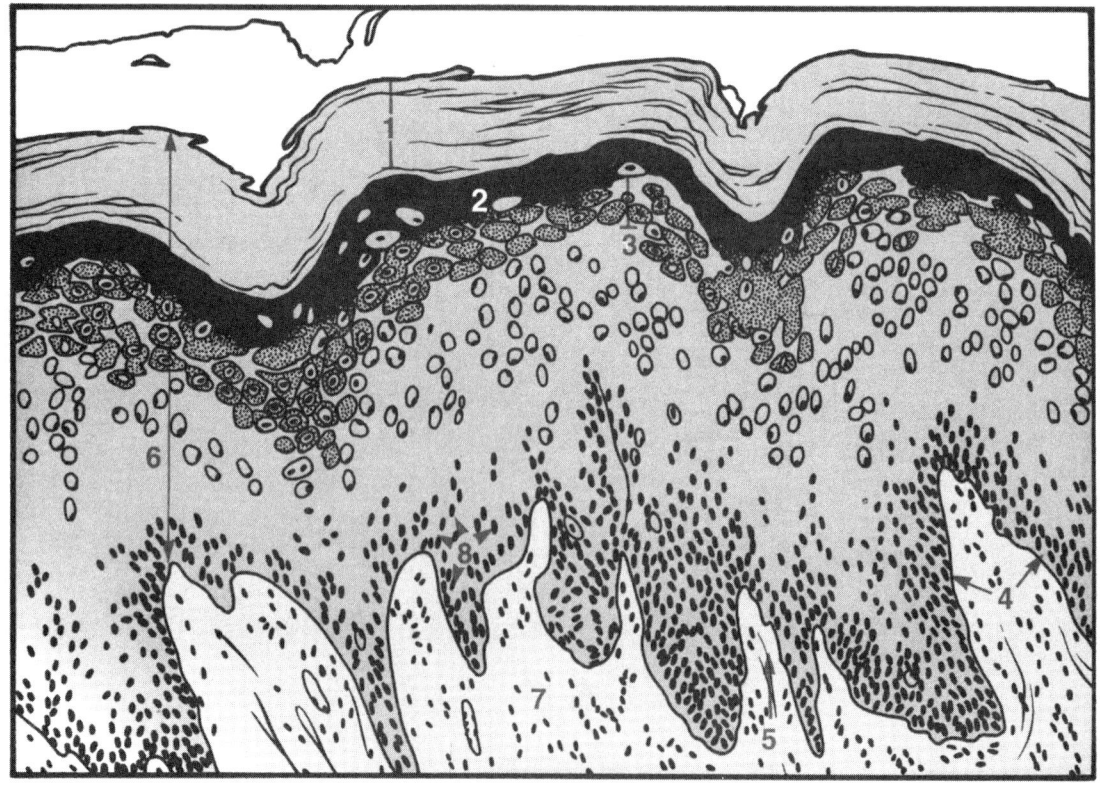

Figure 5.6 Human skull, lateral view

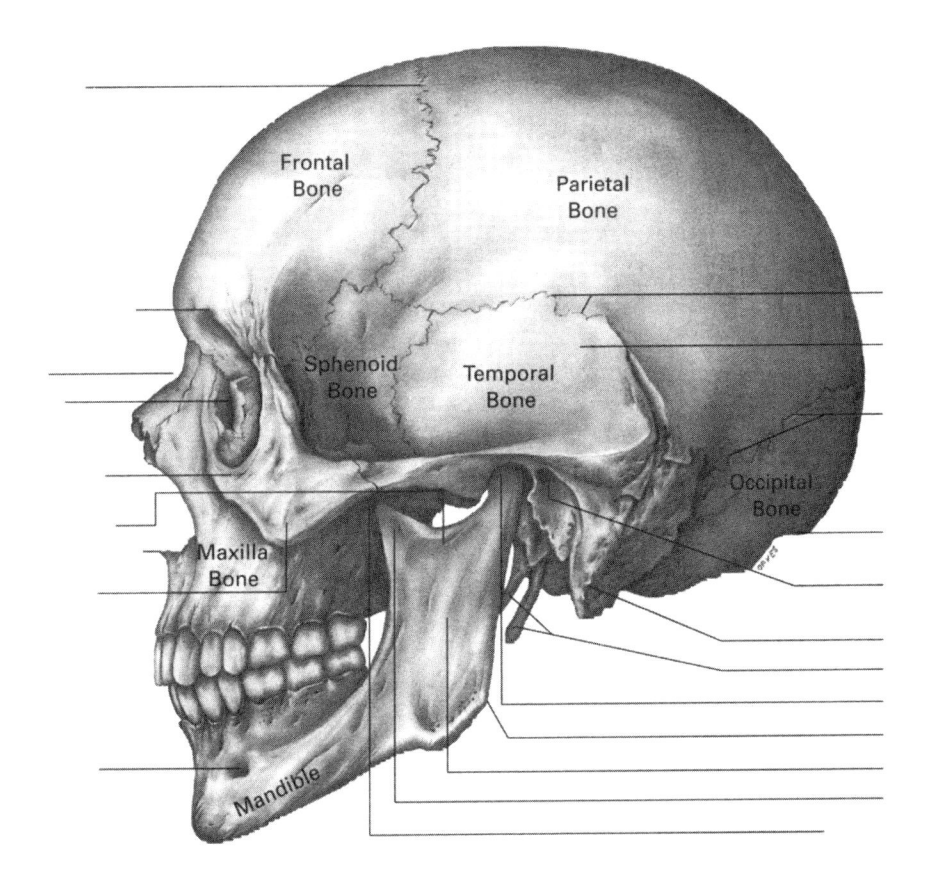

Figure 5.18 Interior aspect of human skull

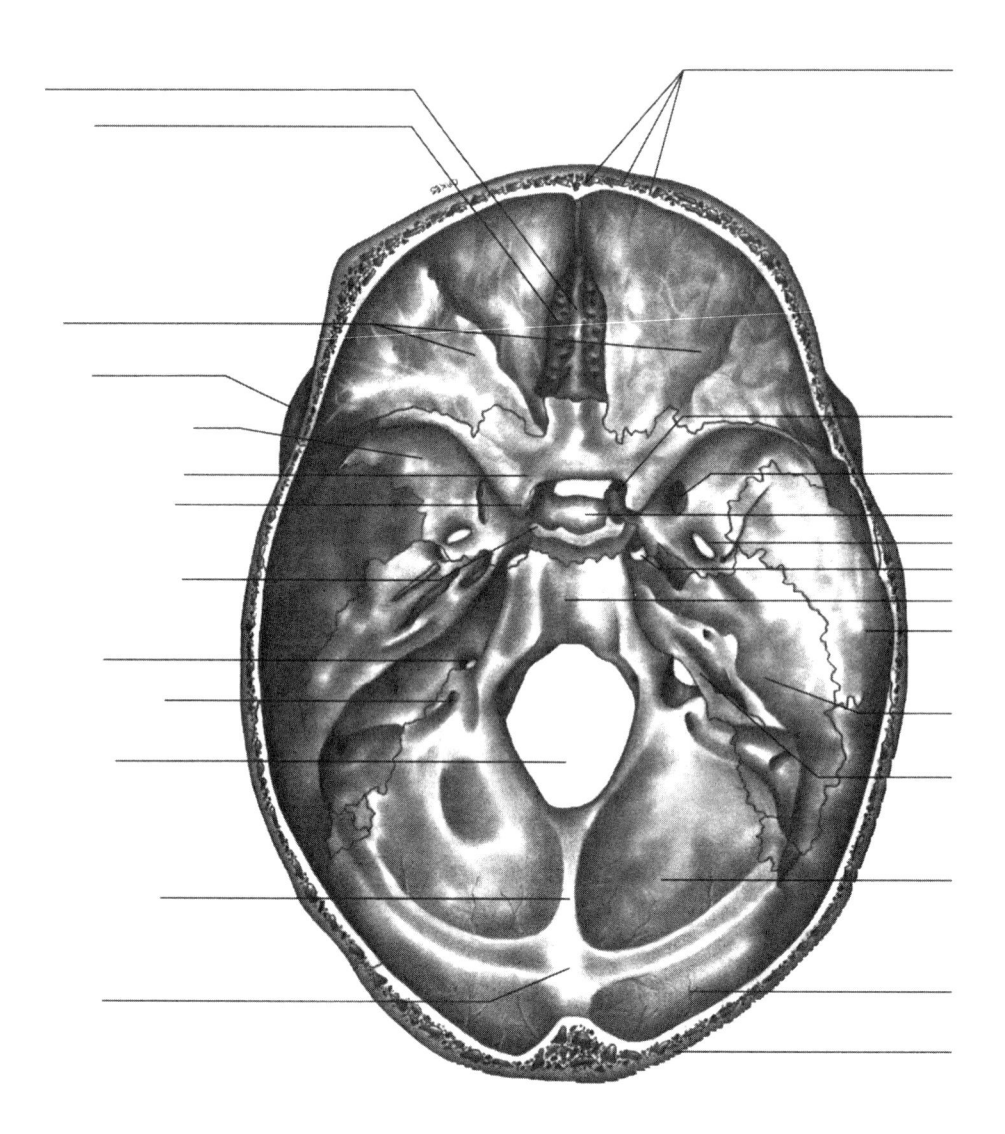

Figure 5.21 Normal curvatures of the human spinal column

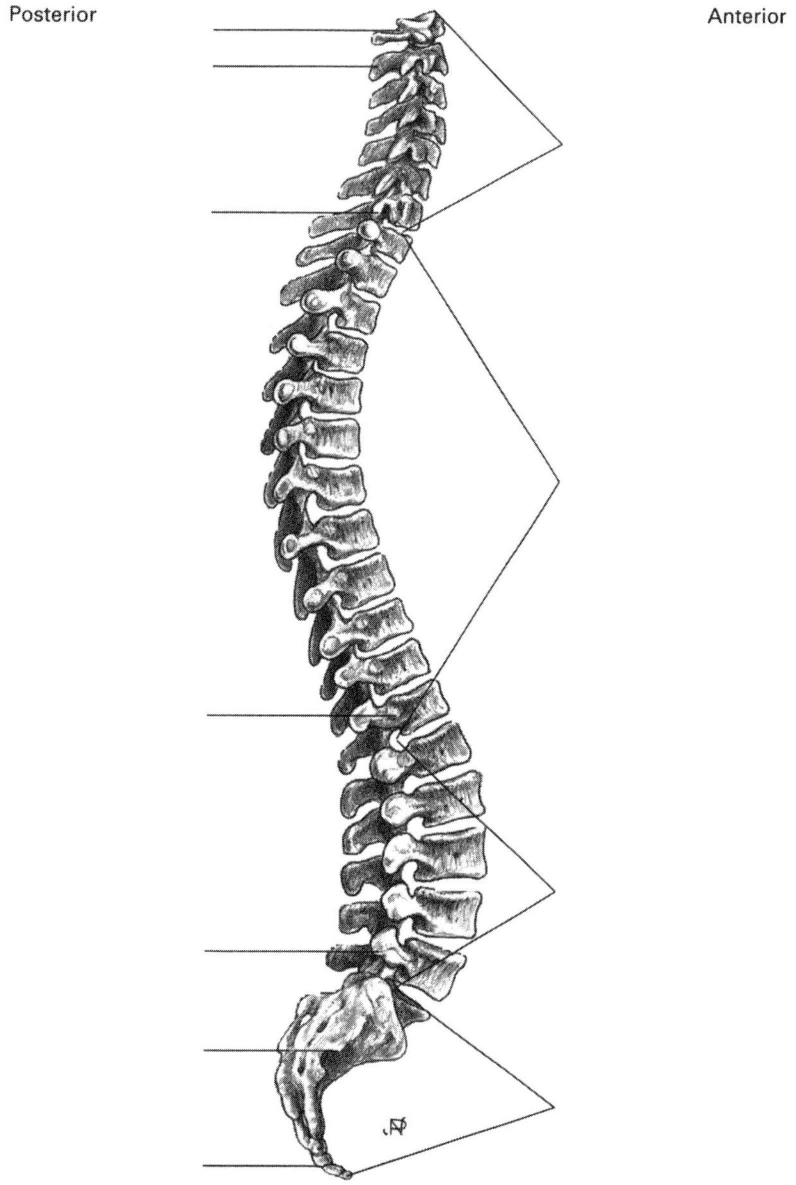

Posterior   Anterior

Figure 5.28 Superior aspect of typical lumbar vertebra

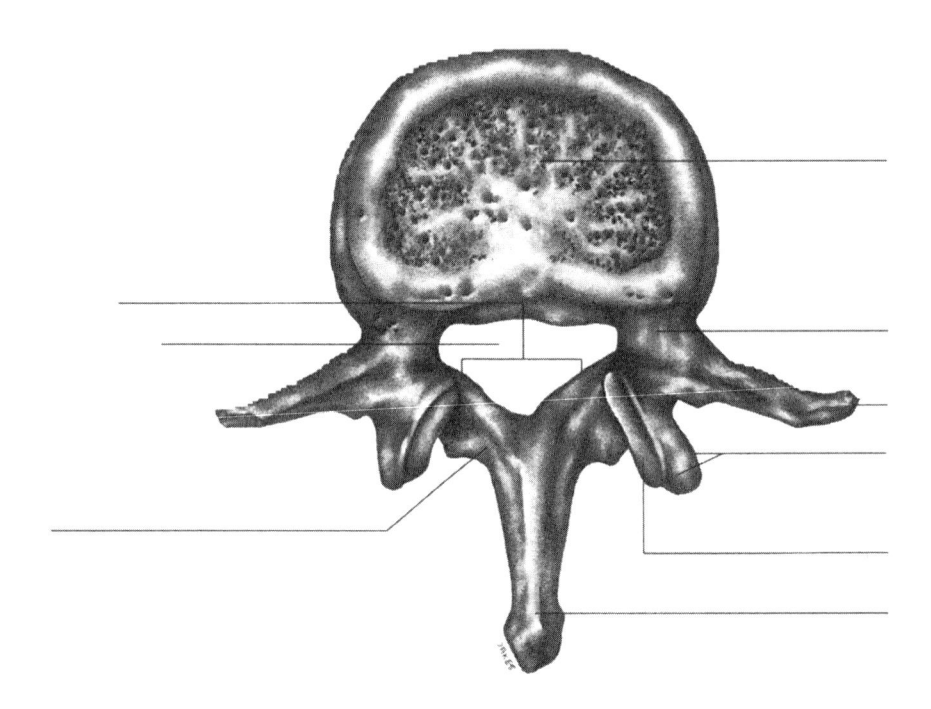

Figure 5.30 Sternum, anterior view

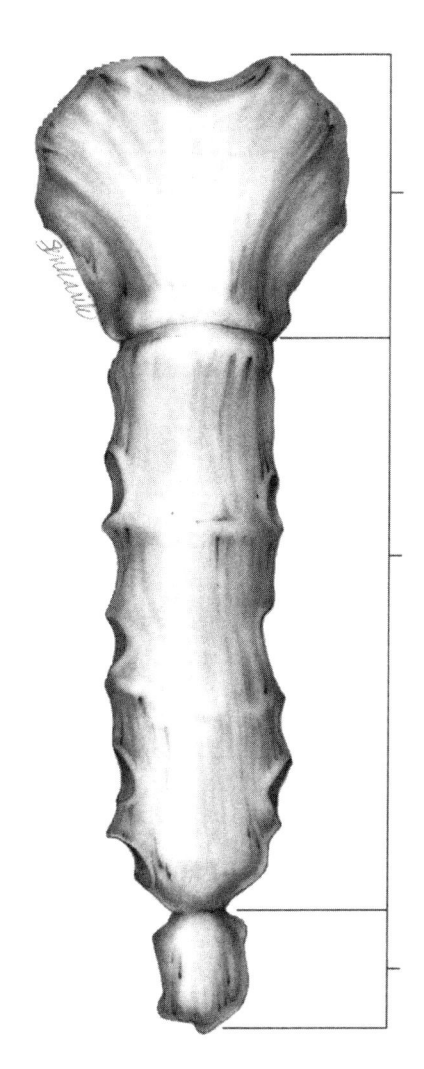

Figure 5.35 Anterior aspect of right humerus

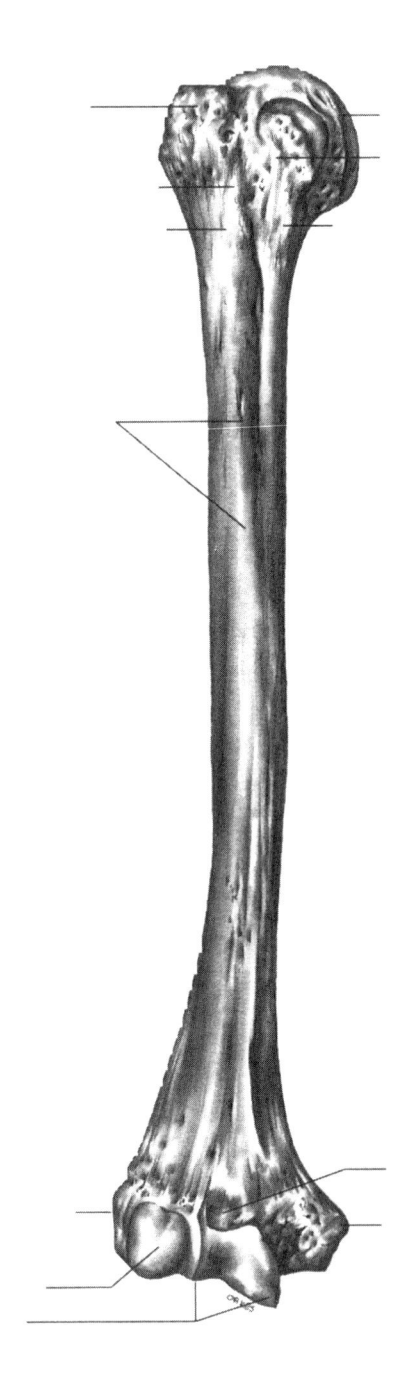

Figure 5.37 Anterior view of right radius and ulna

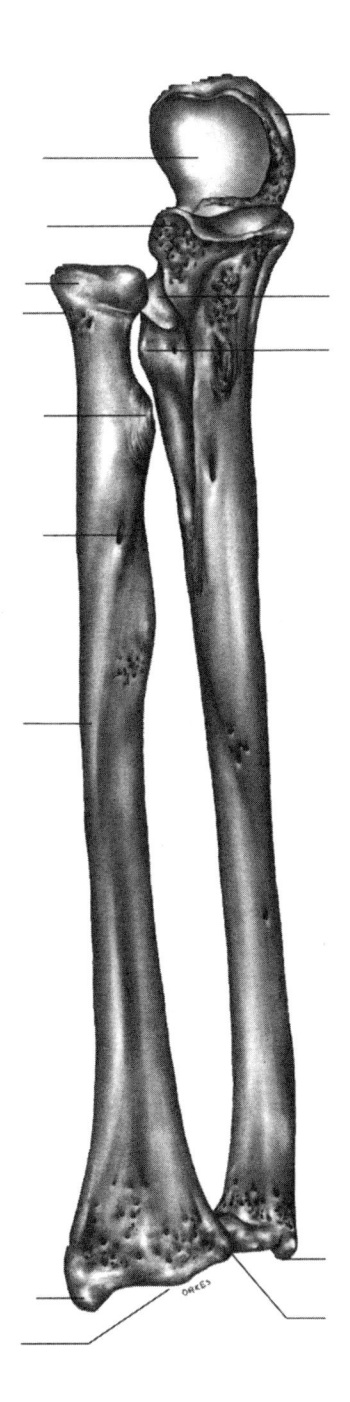

Figure 5.39 Dorsal aspect of bones of right hand

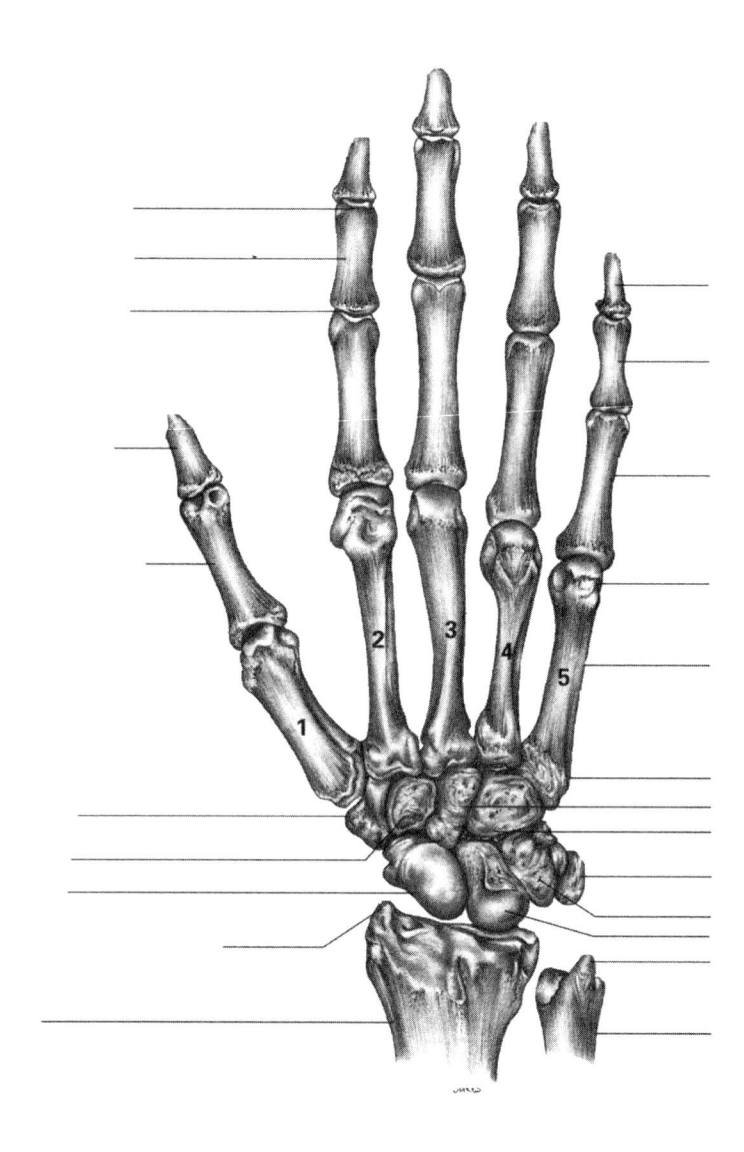

Figure 5.42 Lateral aspect of right os coxa

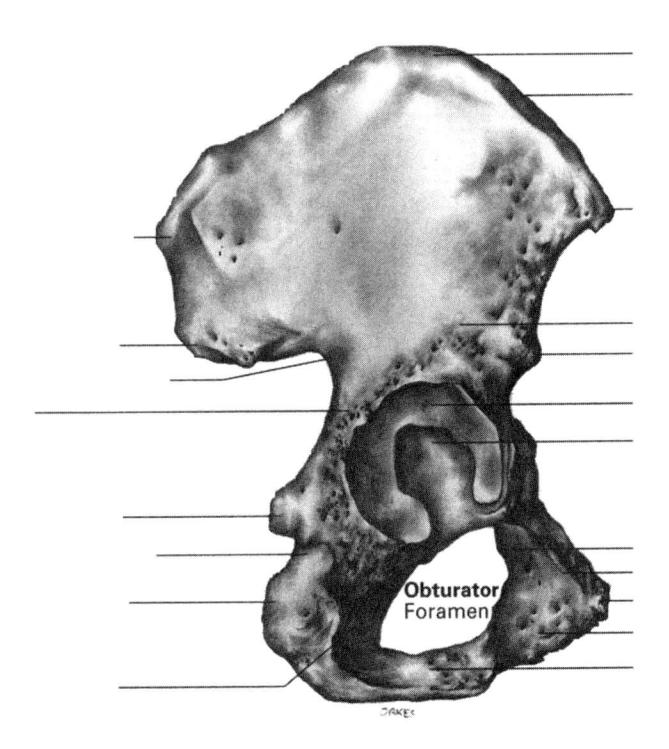

Obturator Foramen

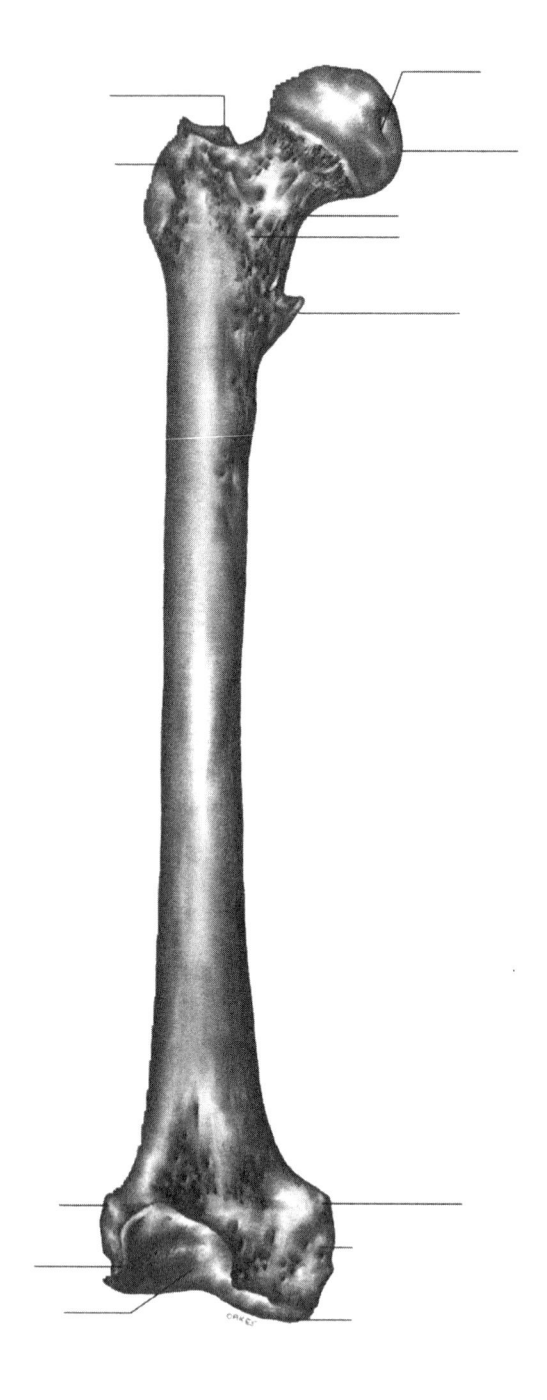

Figure 5.43 Anterior aspect of right femur

Figure 5.47 Anterior aspect of right fibula and tibia

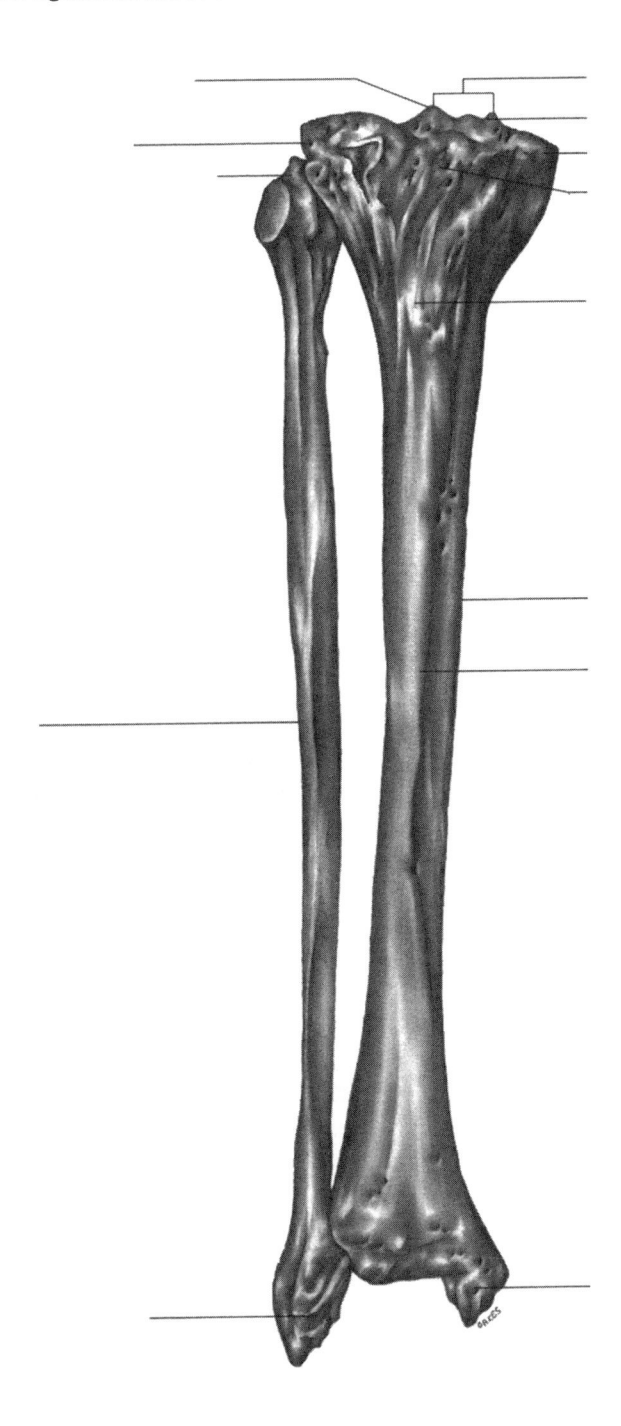

Figure 5.48 Bones of right foot, dorsal surface

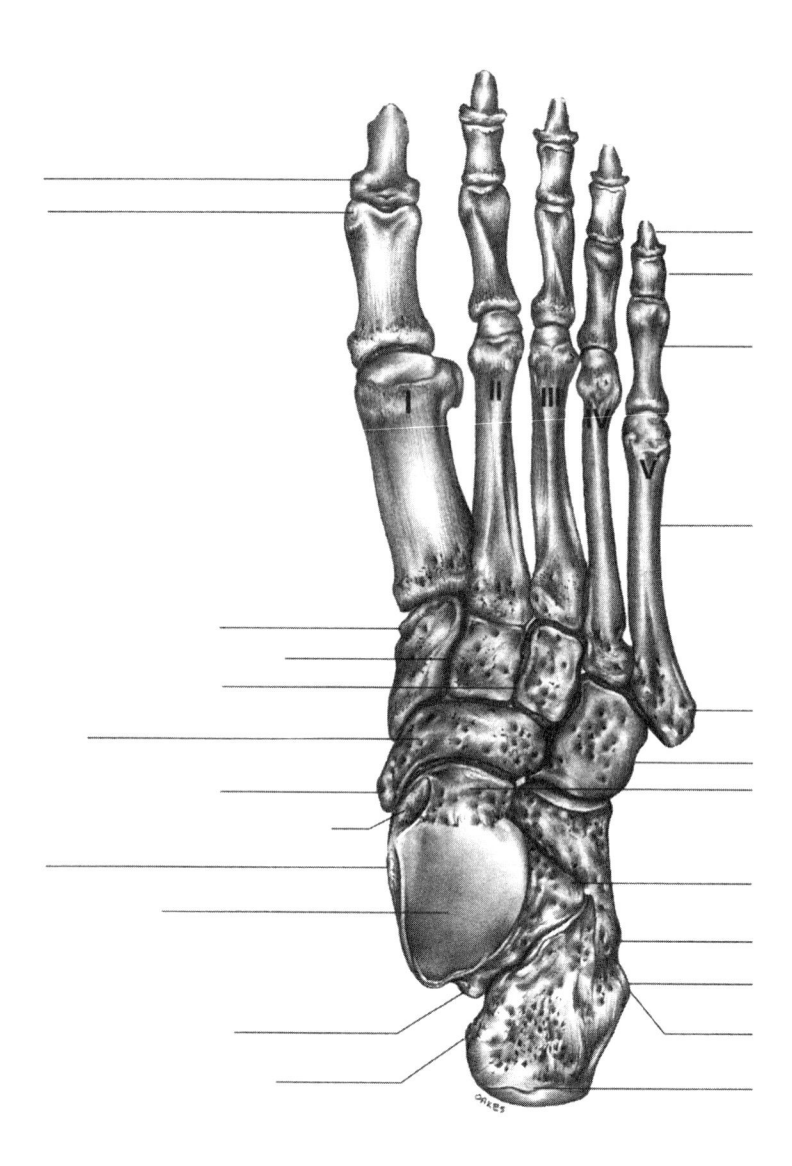

Figure 6.3 Head and neck muscles, anterior view

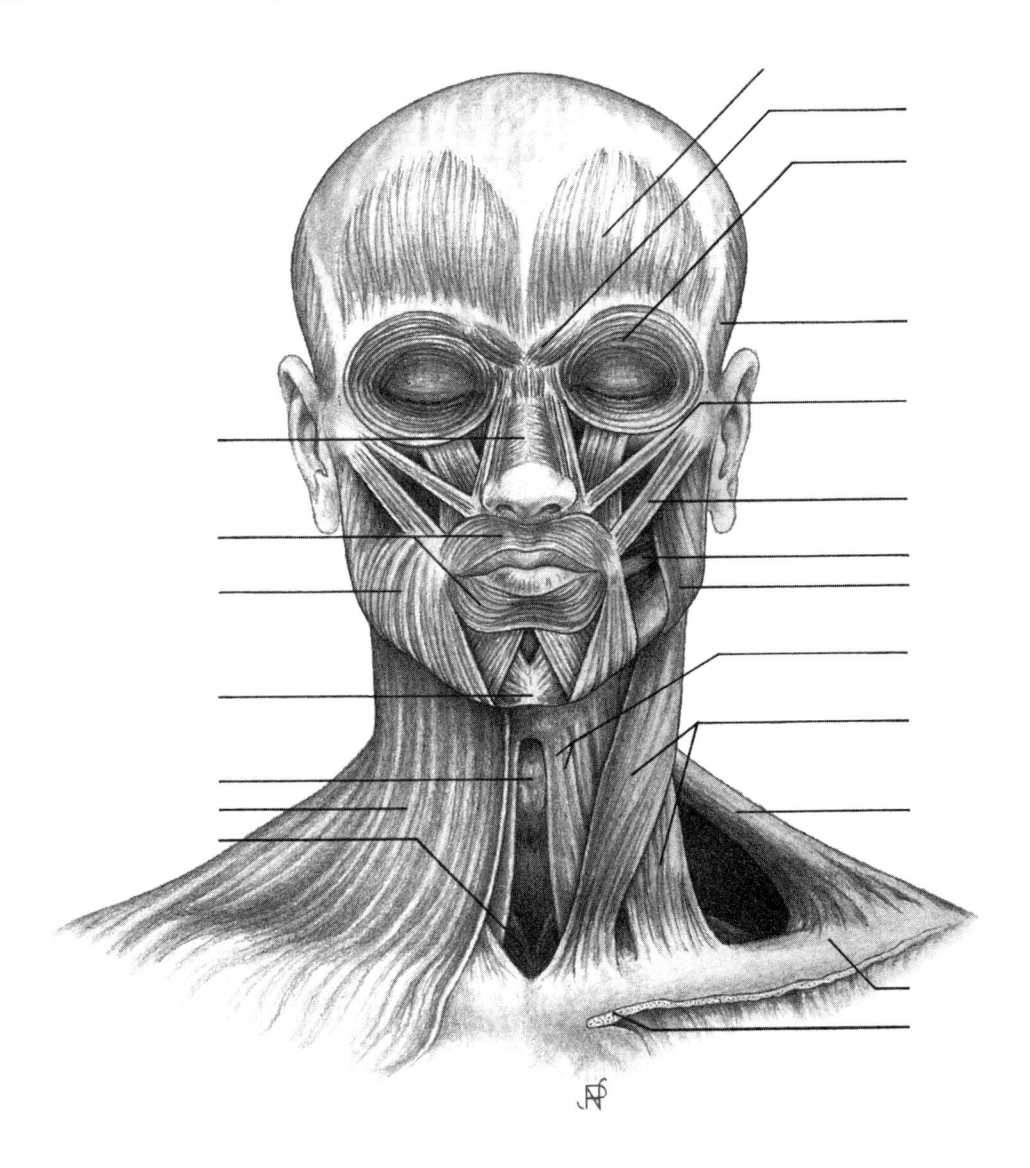

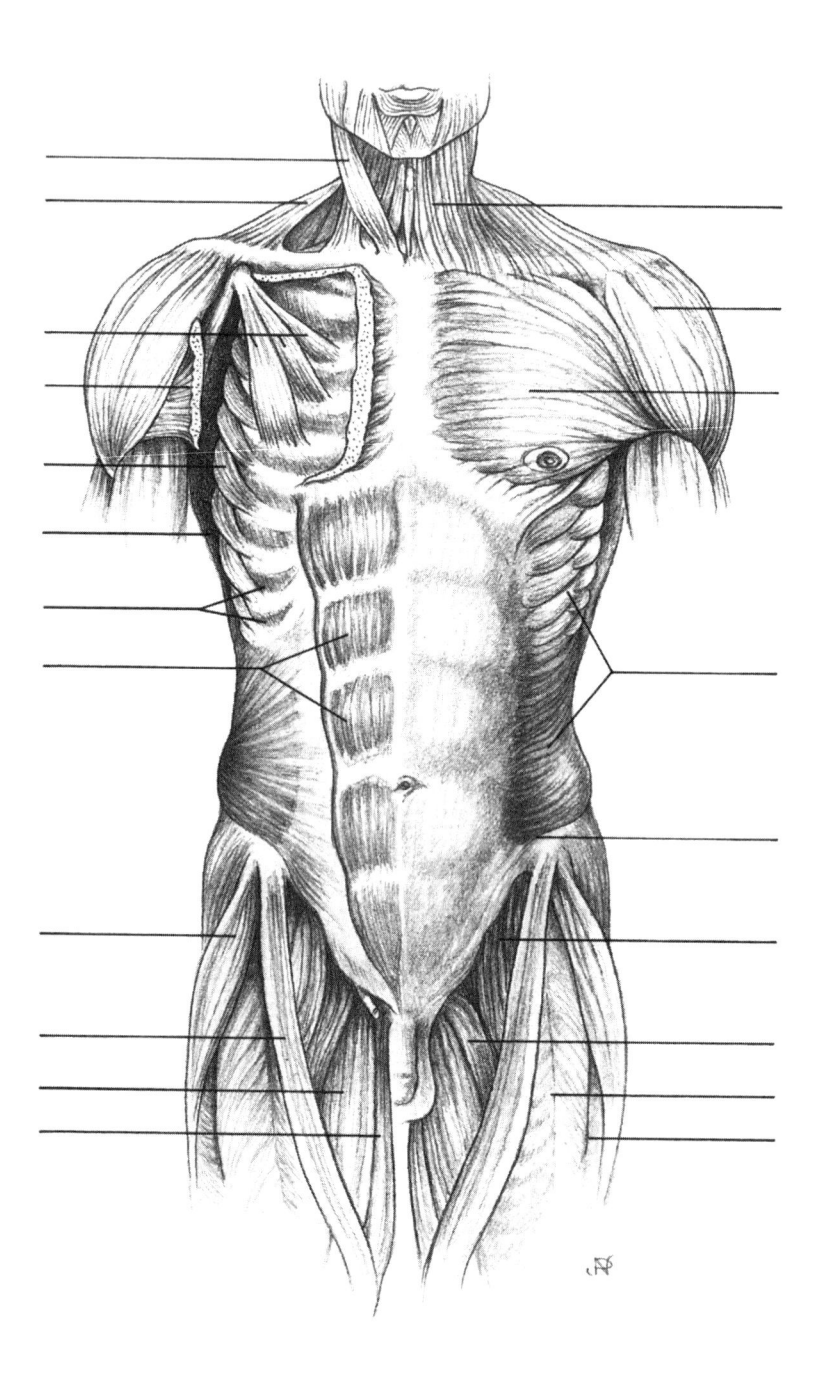

Figure 6.5 Superficial and deep muscles of the trunk, anterior view

Figure 6.6 Superficial muscles of the trunk, posterior view

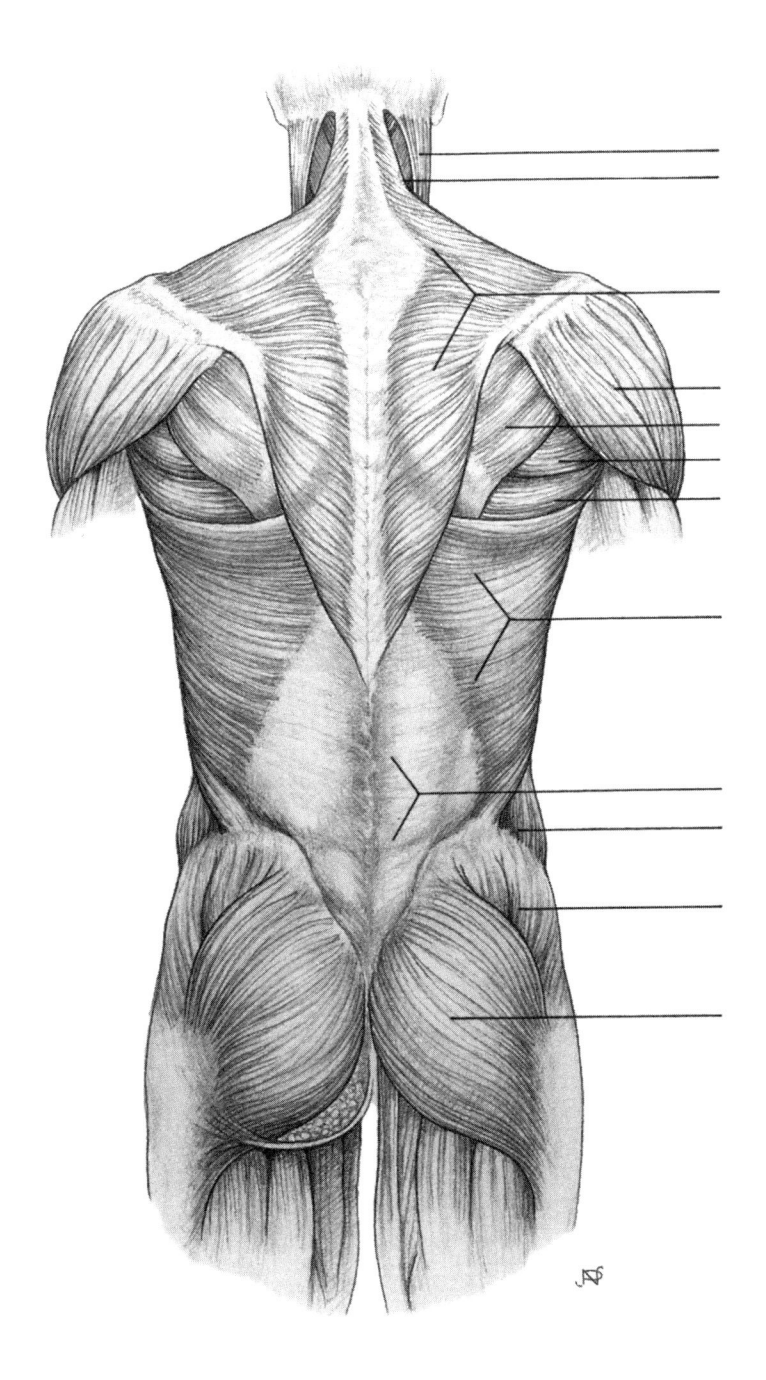

Figure 6.8 Superficial arm and forearm muscles, anatomical position

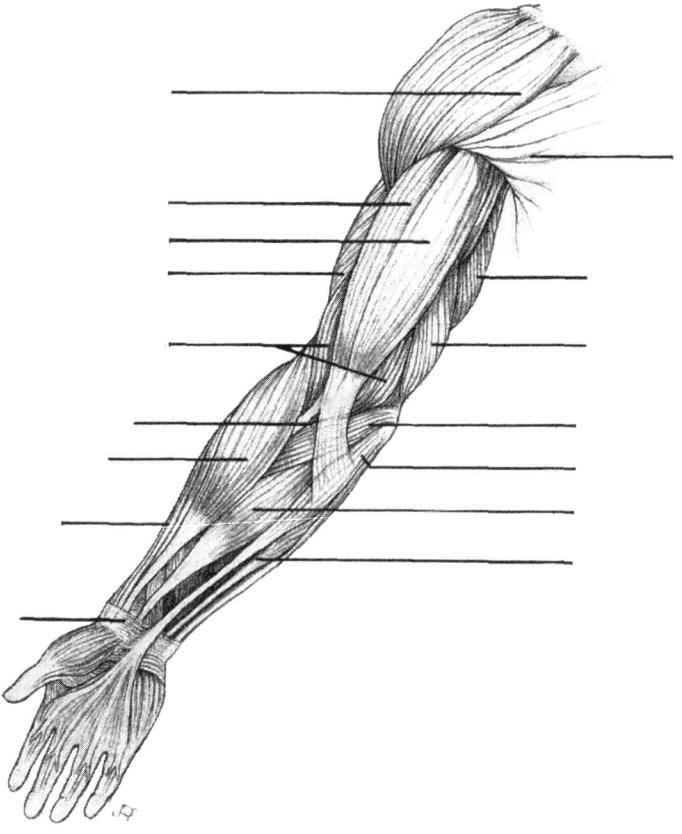

Figure 6.9 Superficial arm and forearm muscles, posterior view

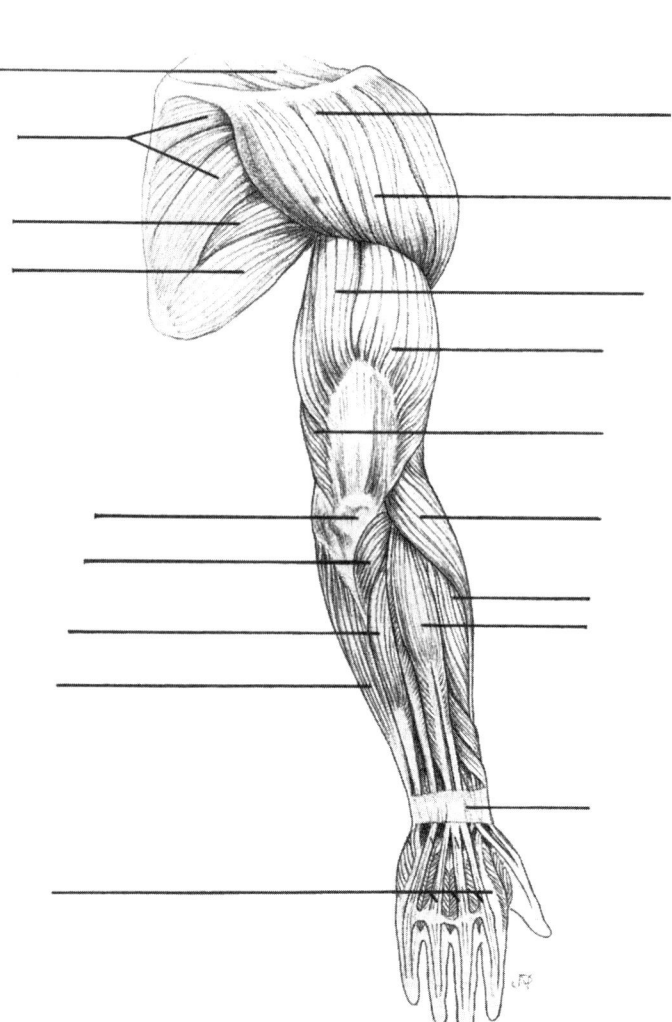

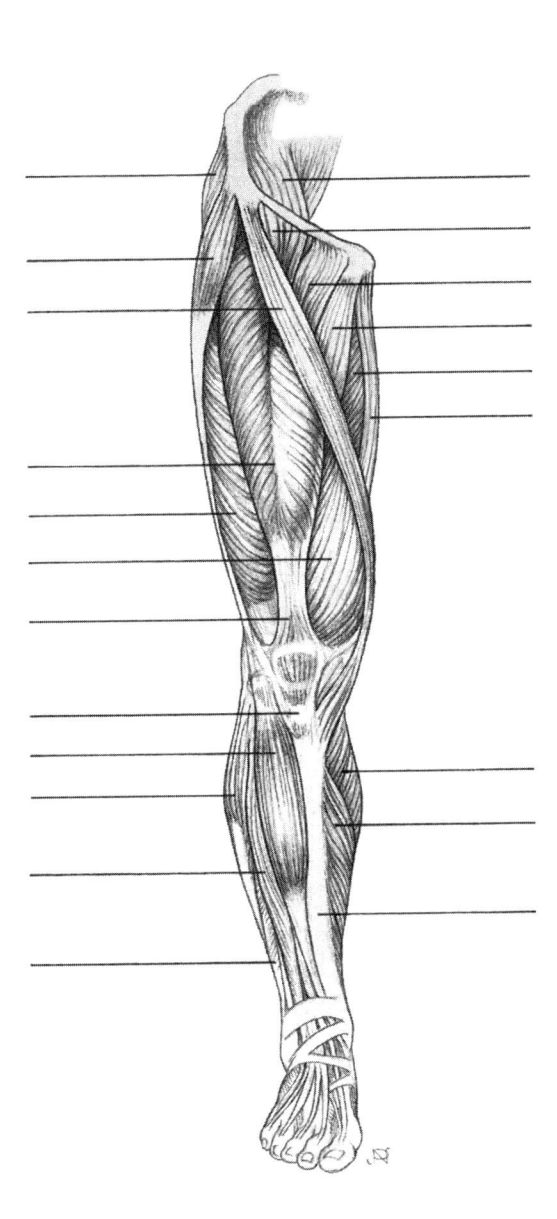

Figure 6.11 Superficial muscles of hip and leg, anterior view

Figure 6.13 Superficial muscles of hip and leg, posterior view

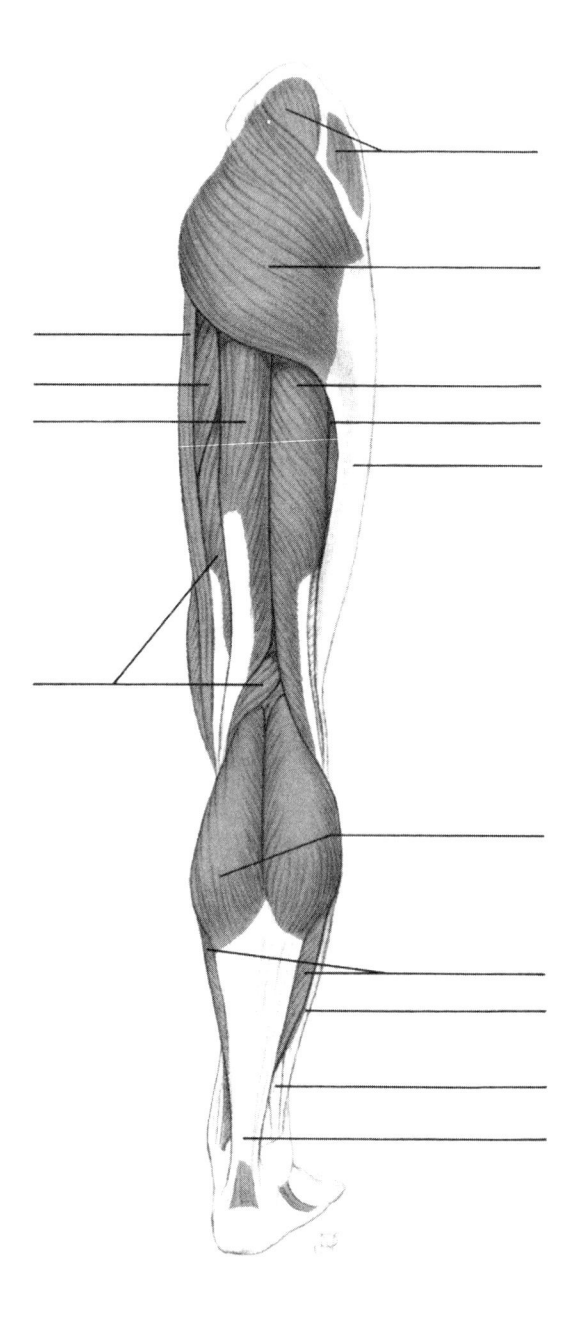

Figure 7.10 Human brain, midsagittal section

Figure 7.16 Human brain, inferior aspect, with cranial nerves

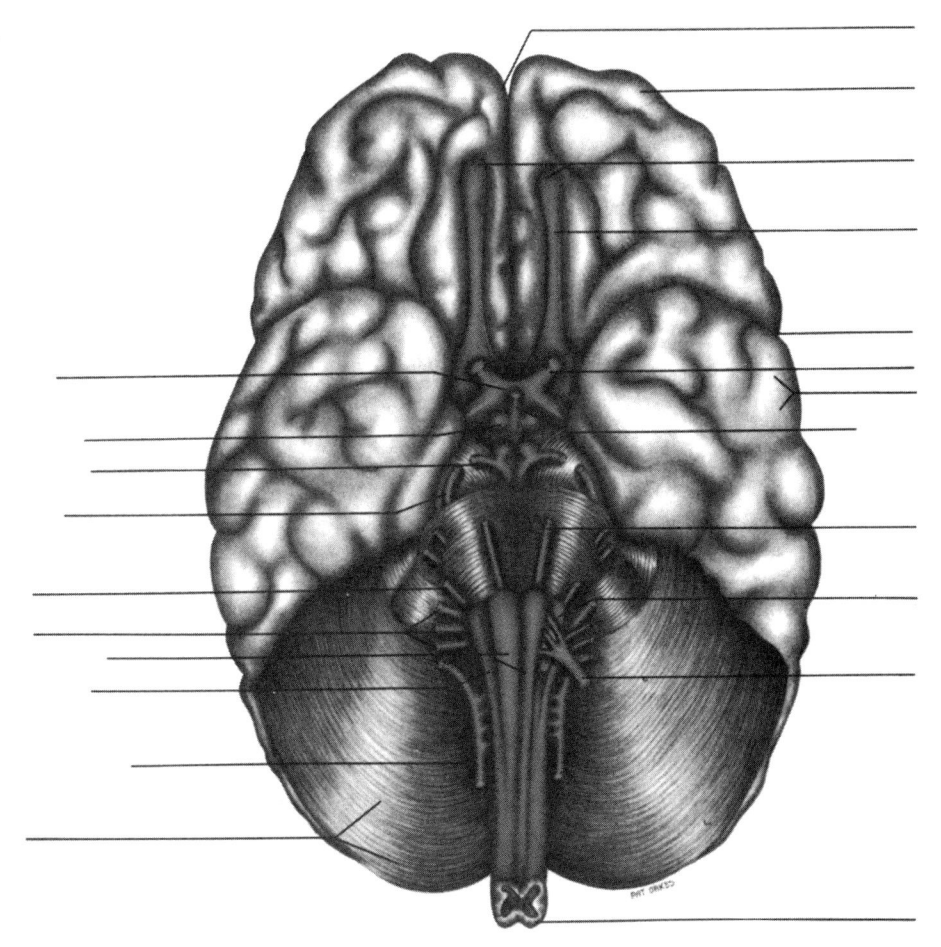

Figure 7.18 A reflex arc

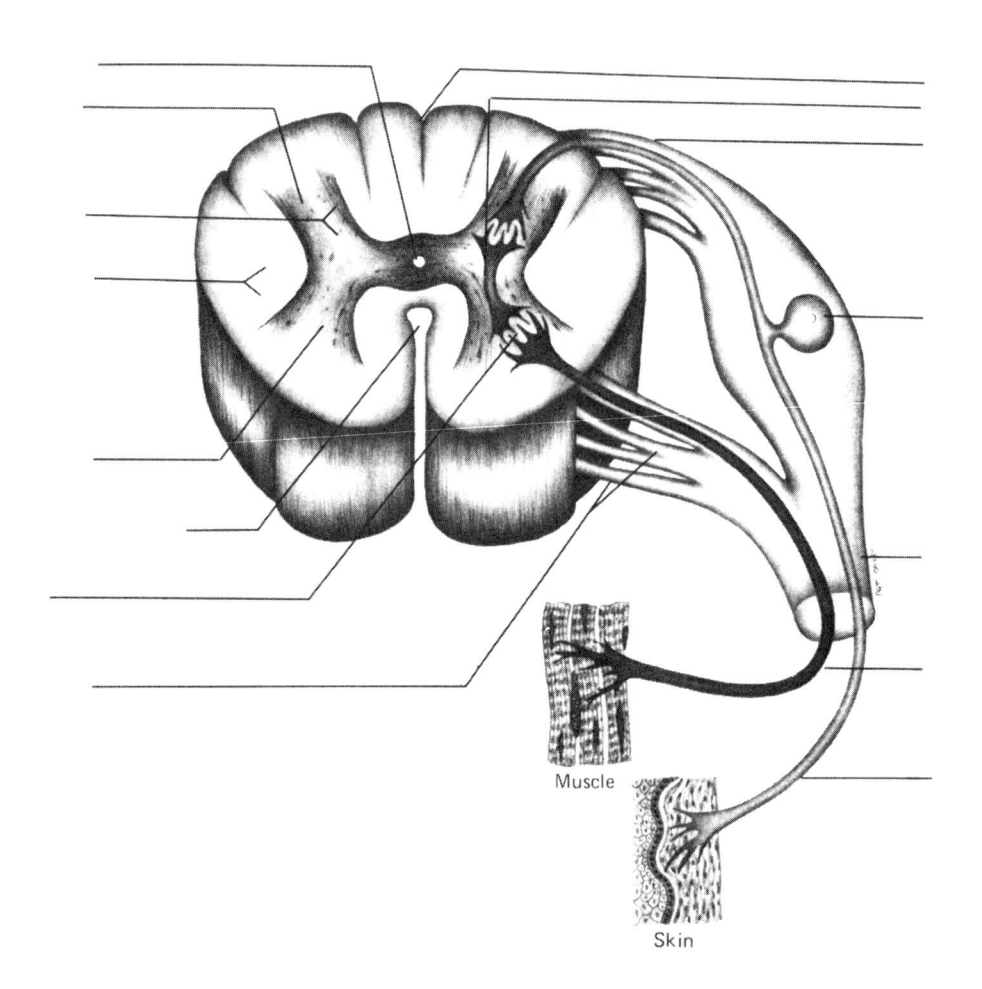

Figure 8.1 Human eye, midsagittal section

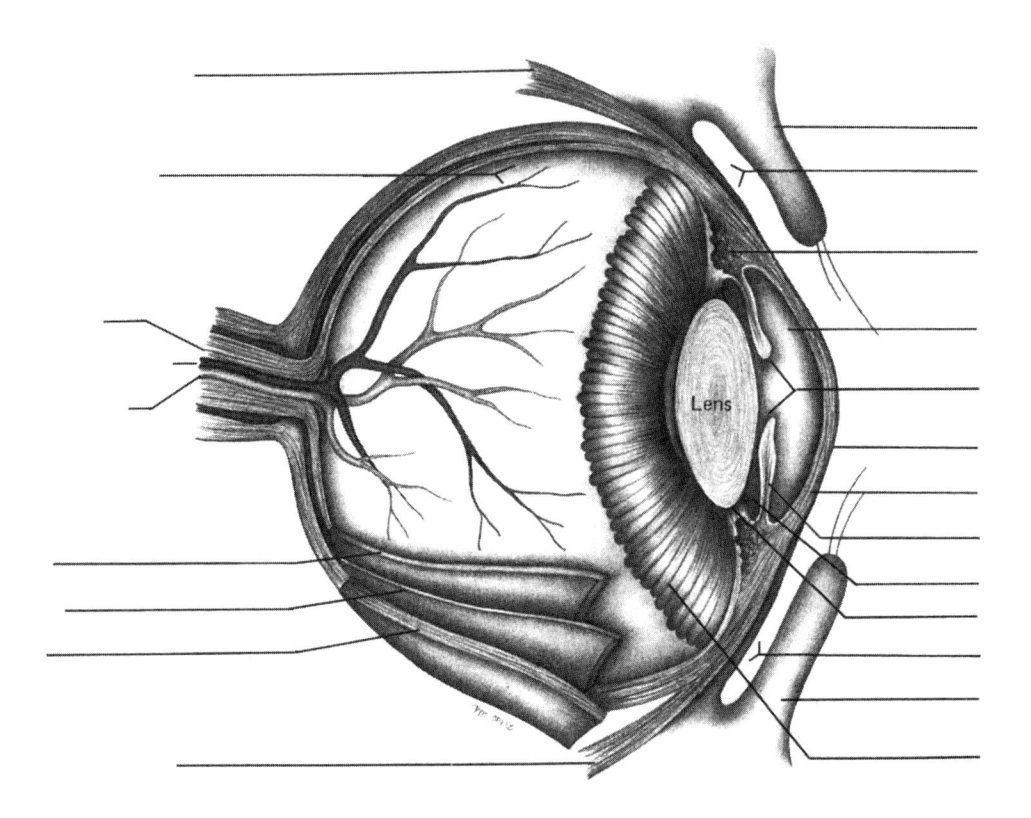

Figure 8.8 Frontal section of right ear showing external, middle, and internal structures

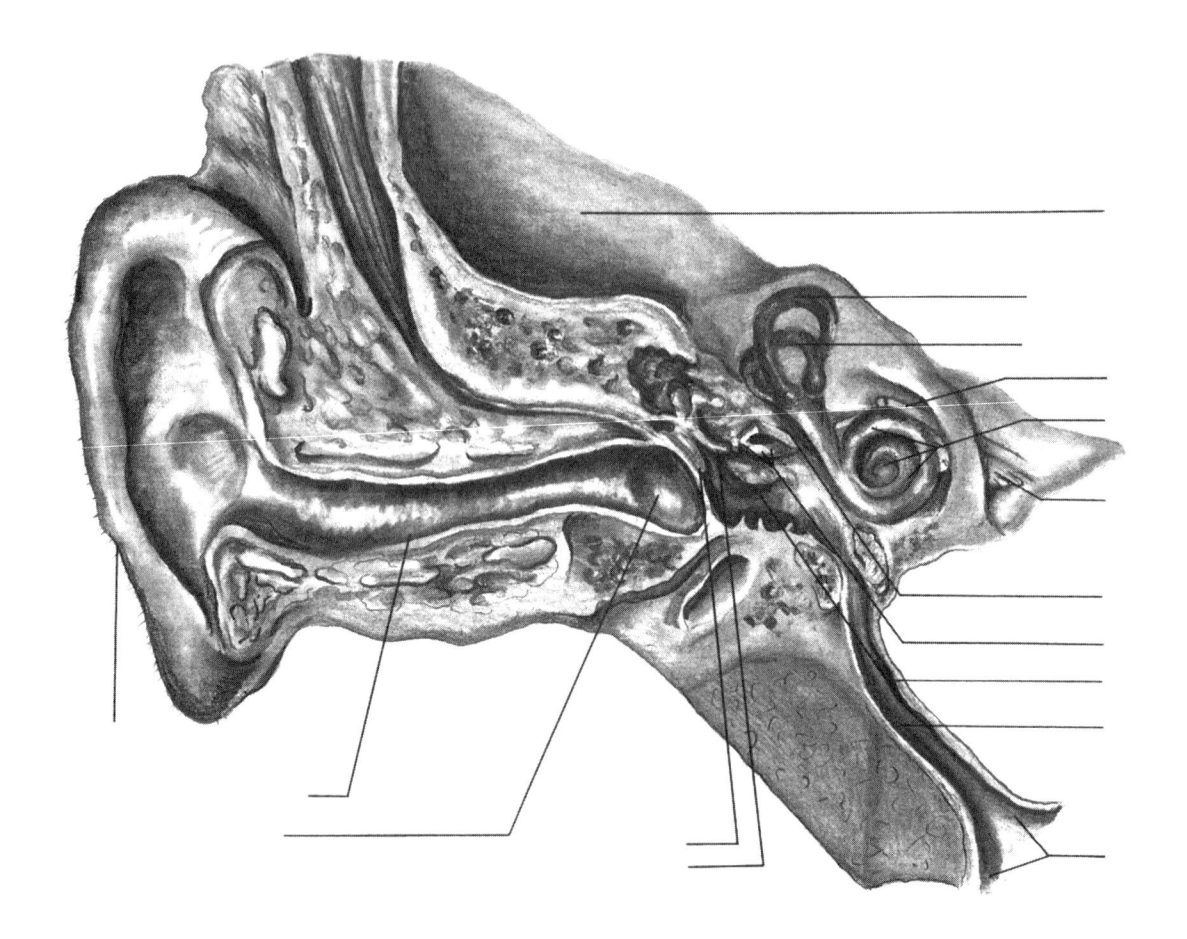

Figure 9A.4 Human lymphatic system

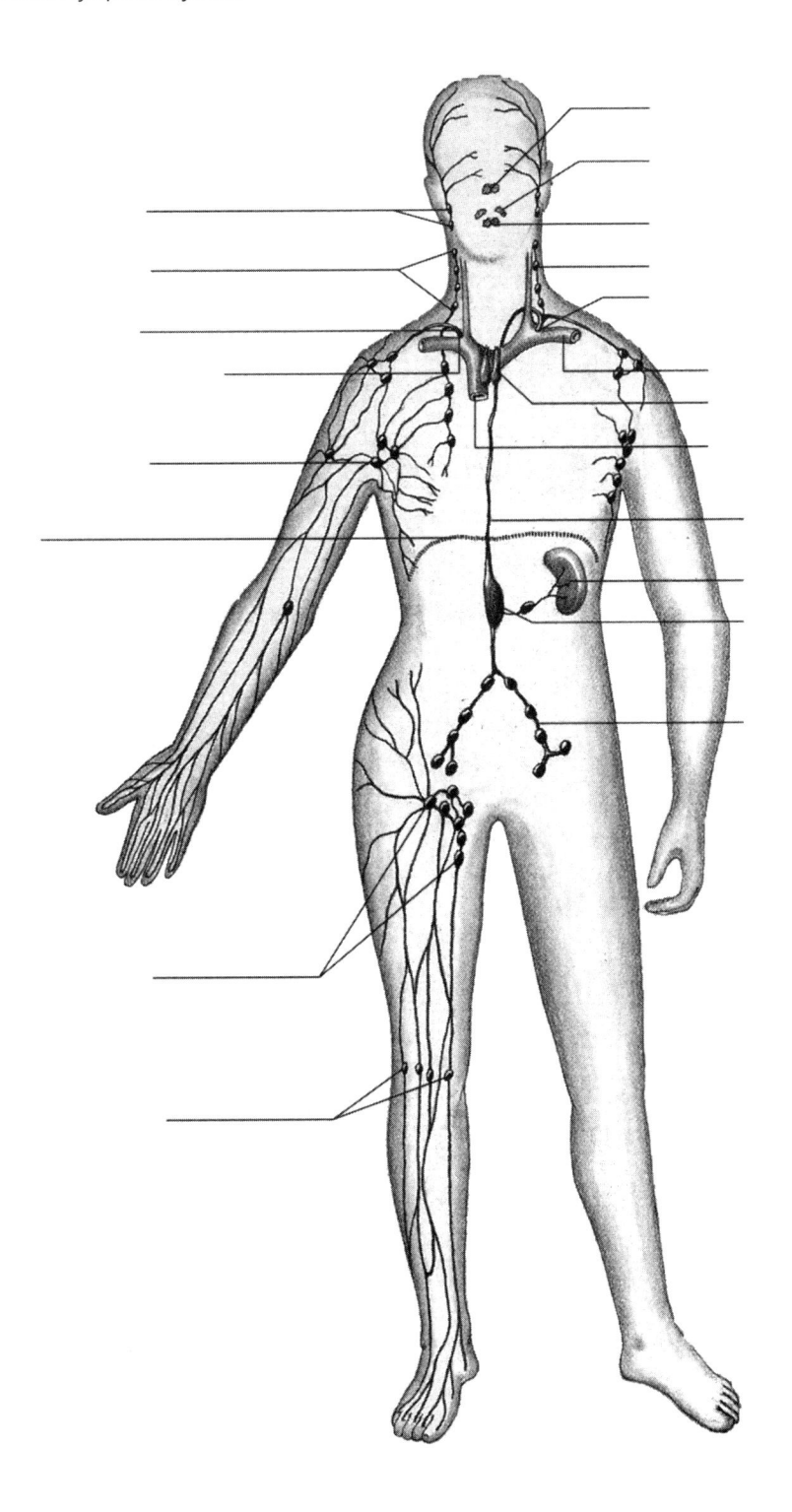

Figure 9.15 Human heart *in situ* with pericardial sac removed, ventral aspect

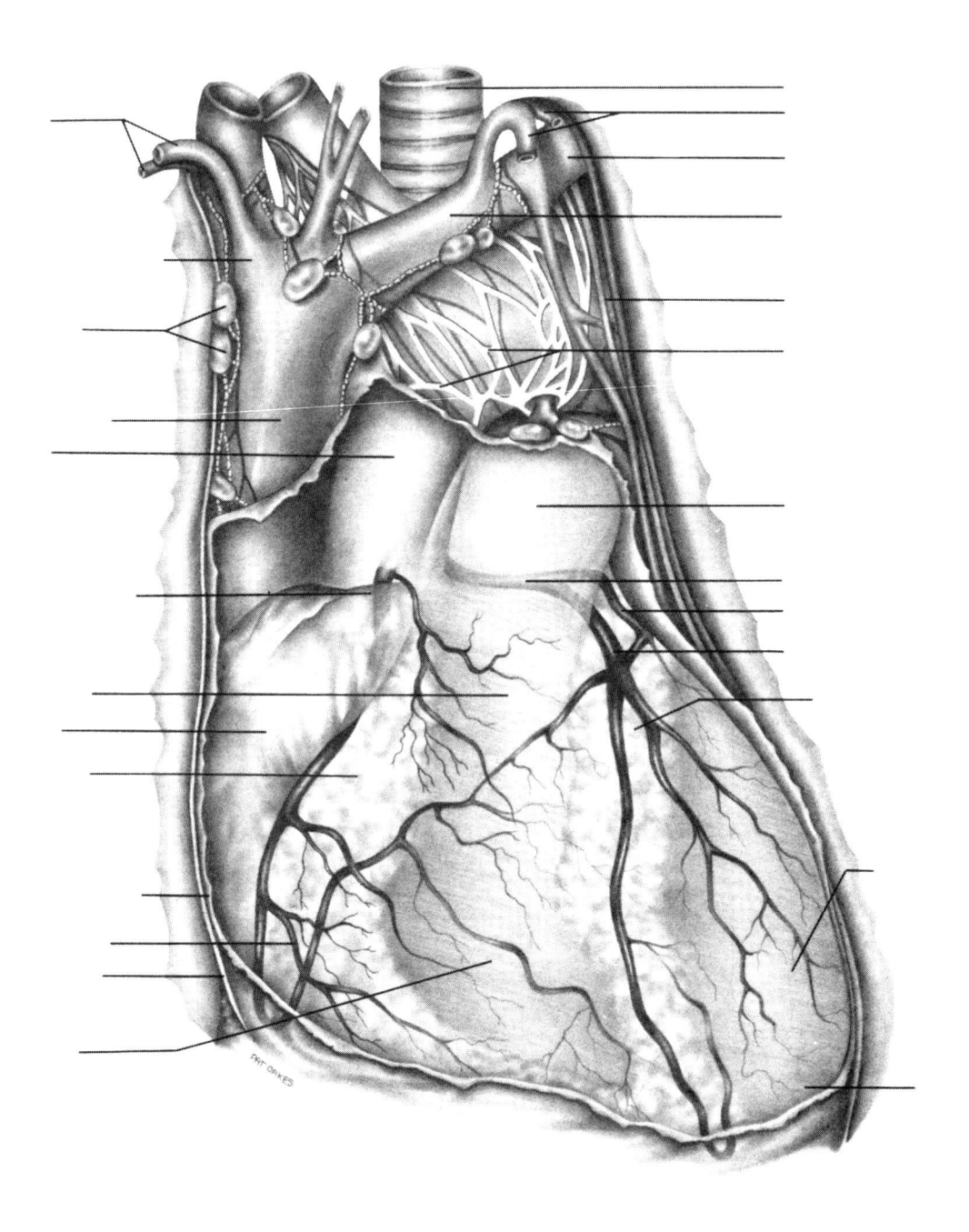

Figure 9A.14 Human heart showing chamber detail, ventral aspect

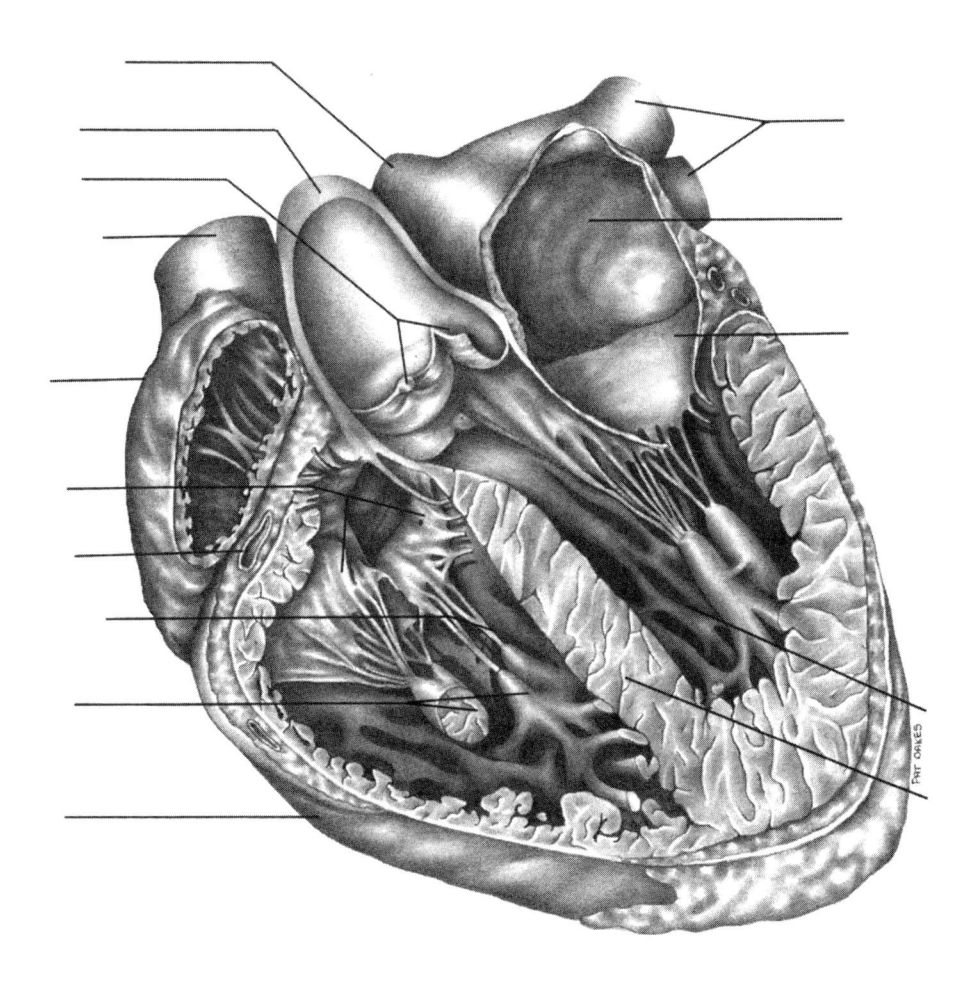

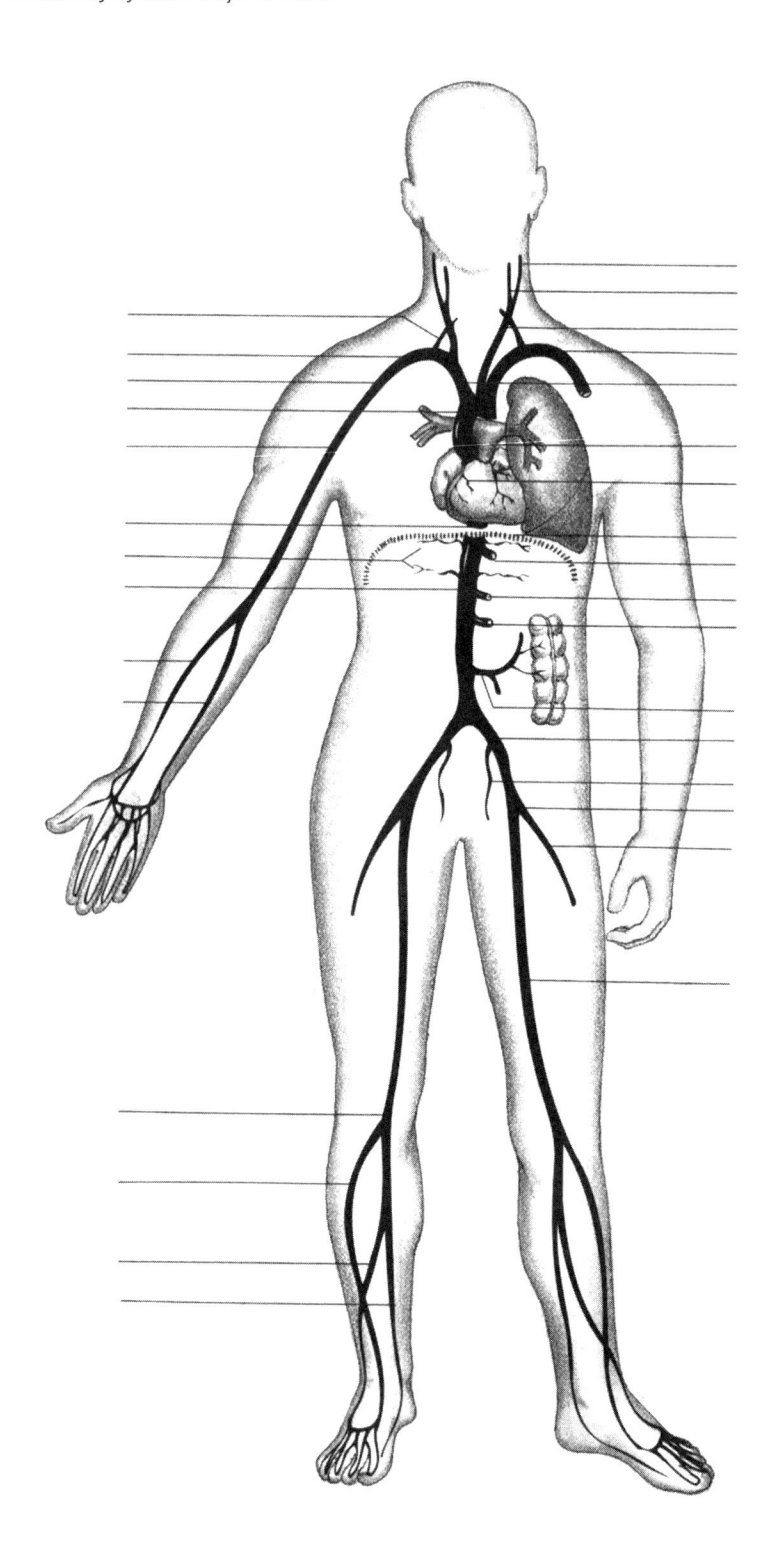

Figure 9.25 Human circulatory system: major arteries

Figure 9.26 Human circulatory system: major veins

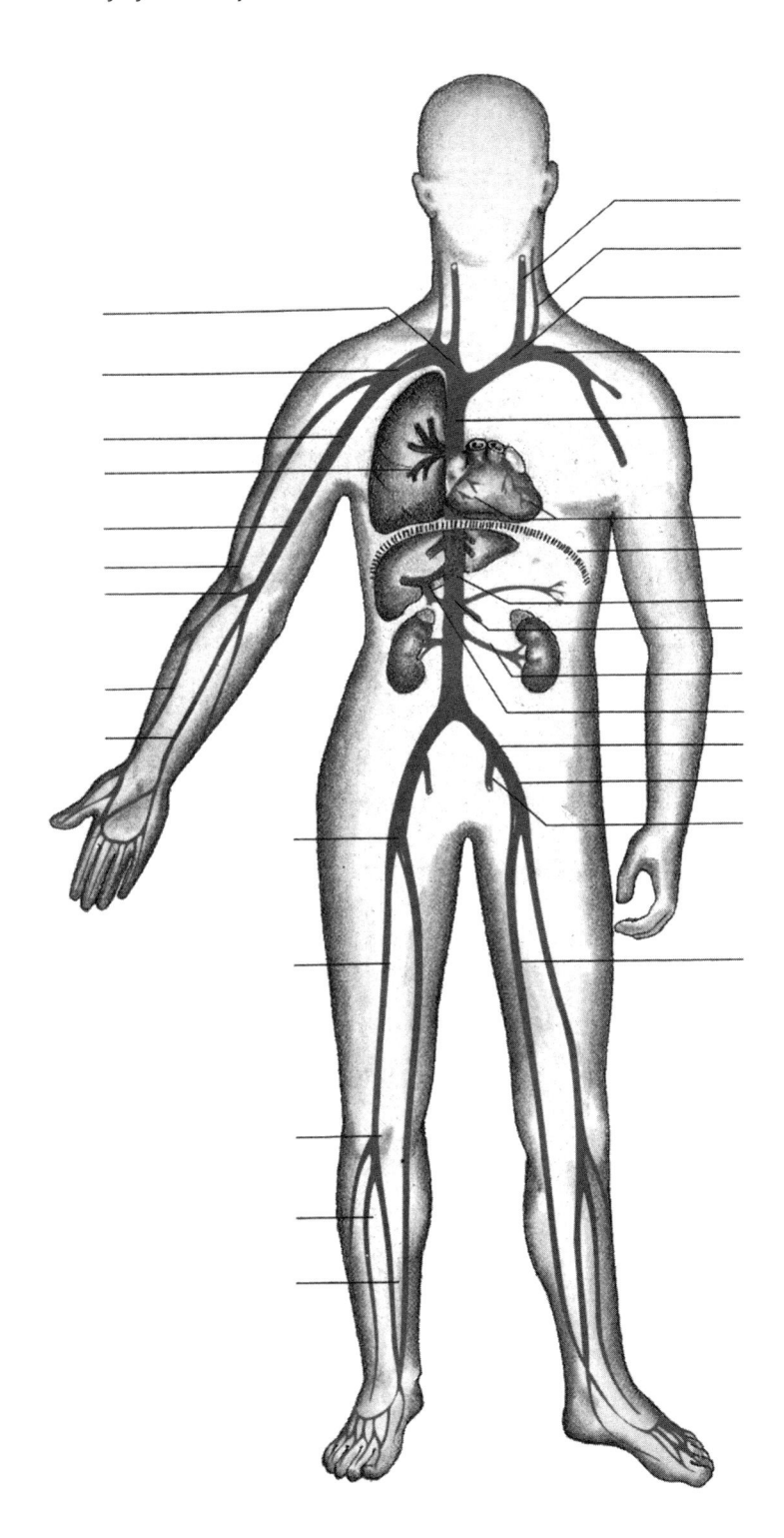

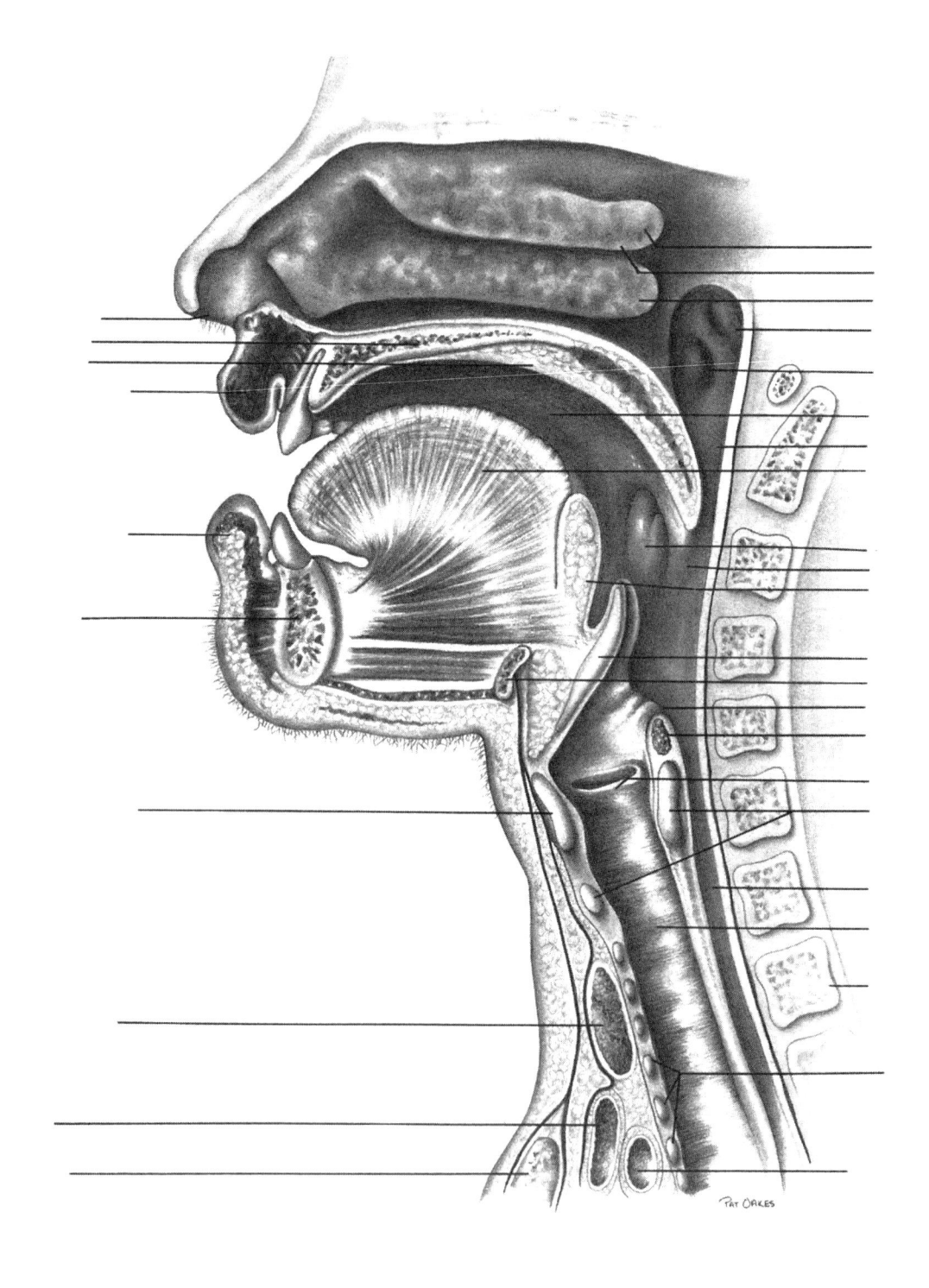

Figure 10.5 Sagittal section of human head

Figure 10.8 Human lung, sagittal section

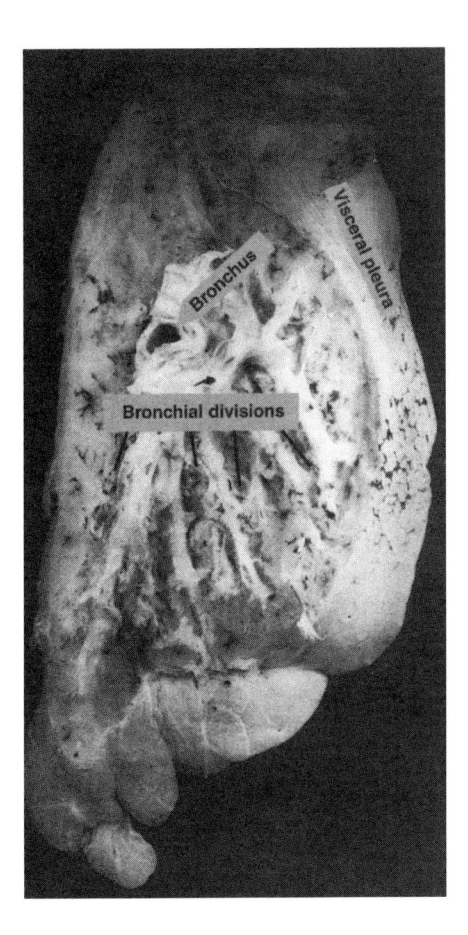

Figure 10.9 Human lung, showing termination of bronchiole into alveolus

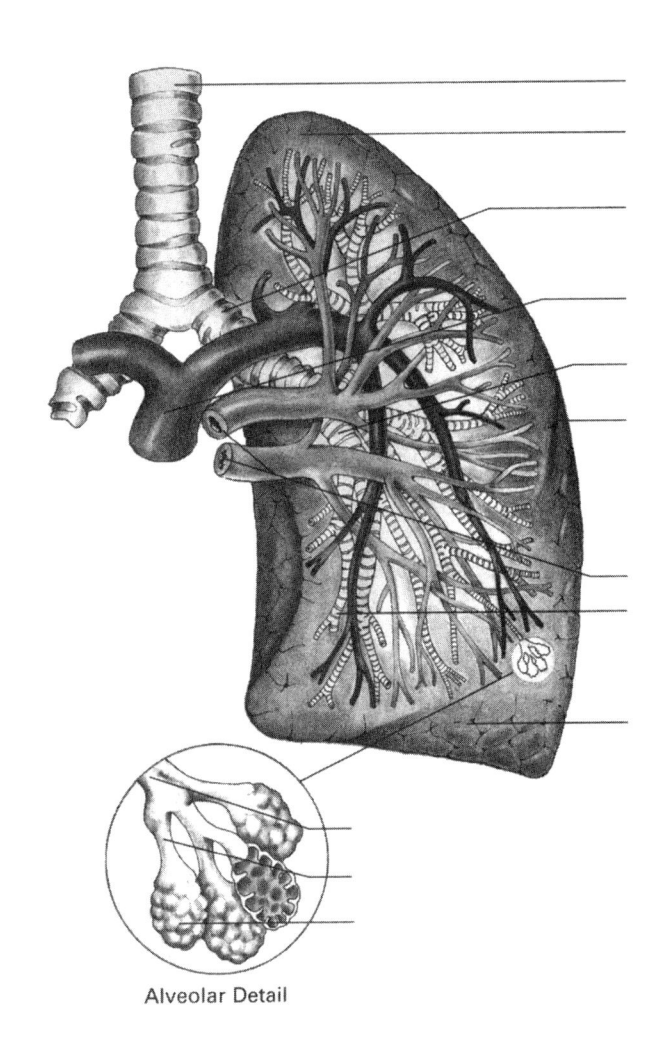

Alveolar Detail

Figure 11.3 Relationship of the liver, gallbladder, and pancreas to the duodenum

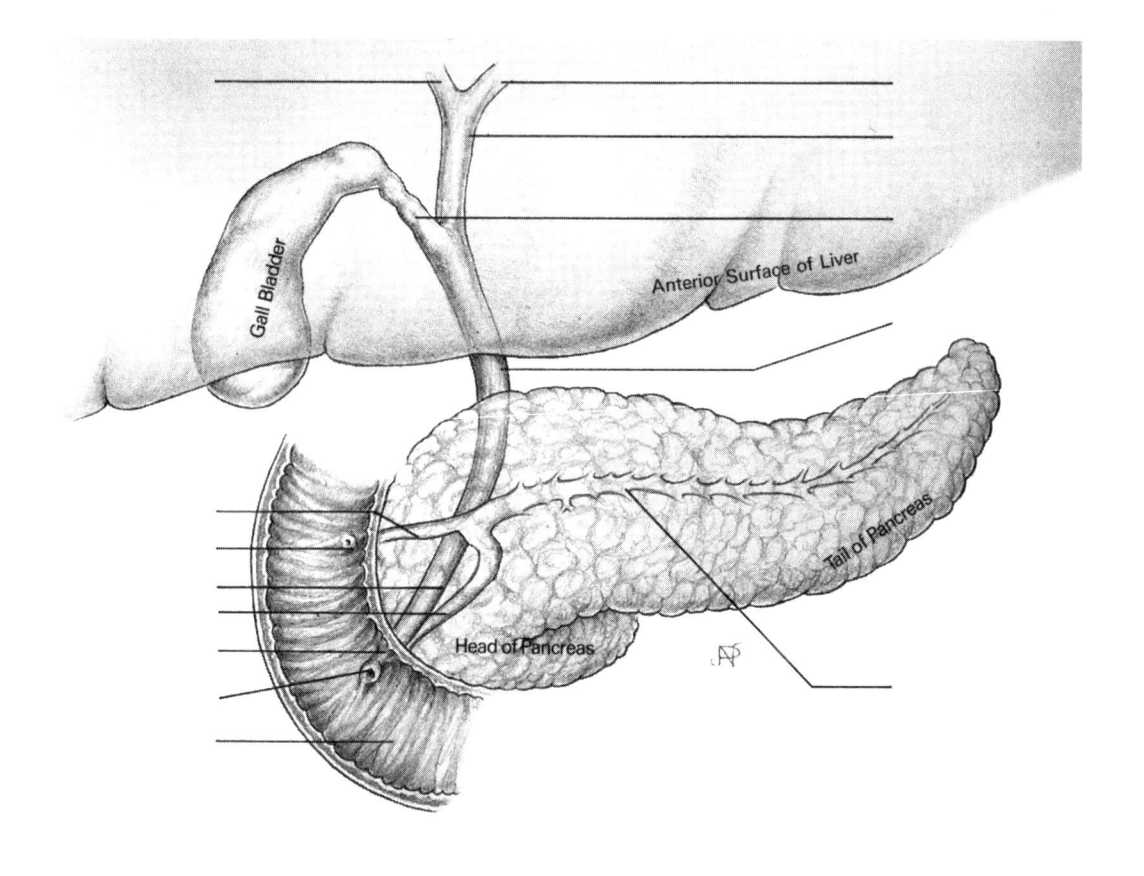

Figure 12A.1 Nephron showing histological detail of tubules

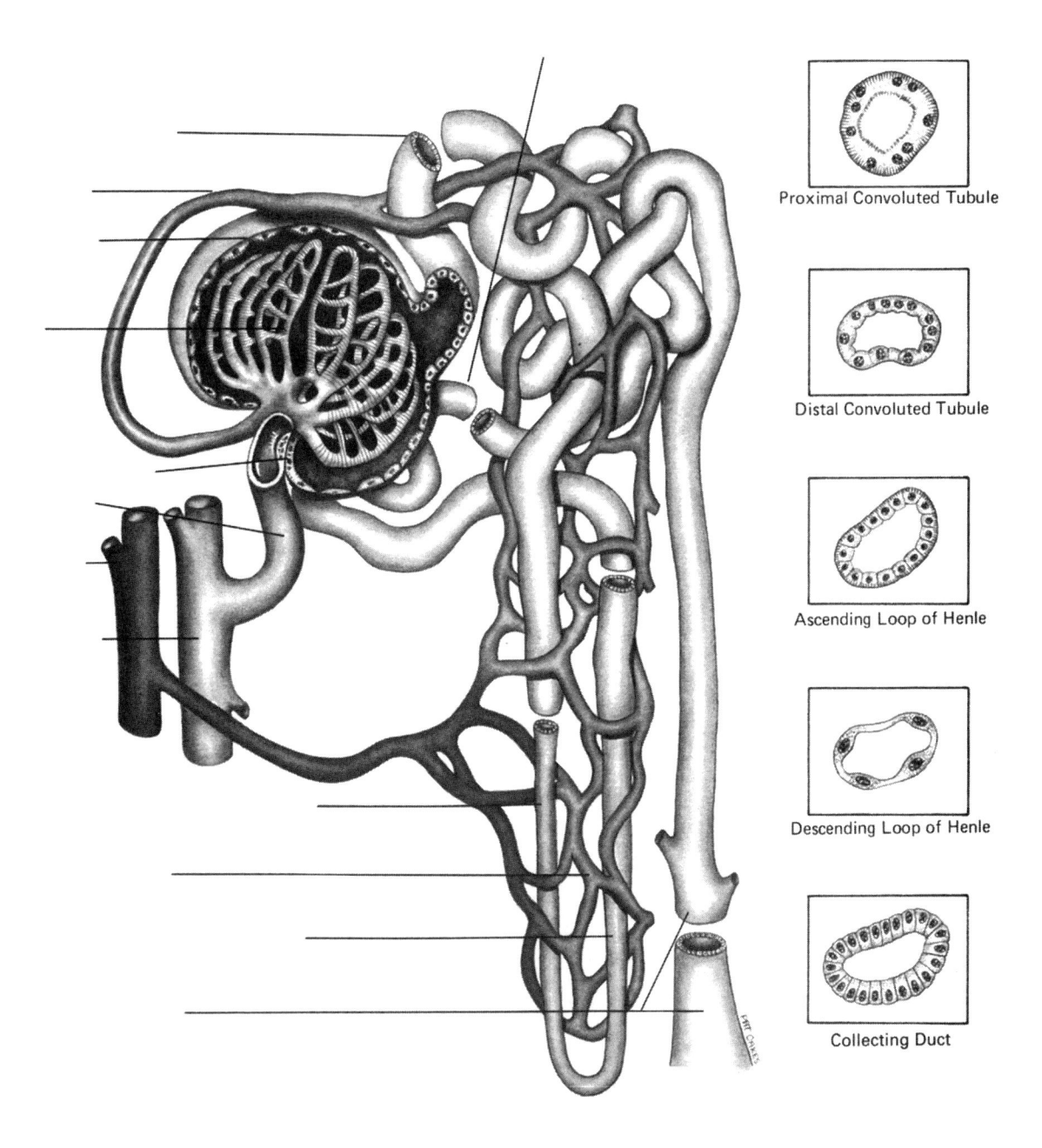

Figure 12.7 Frontal view of human male urinary system

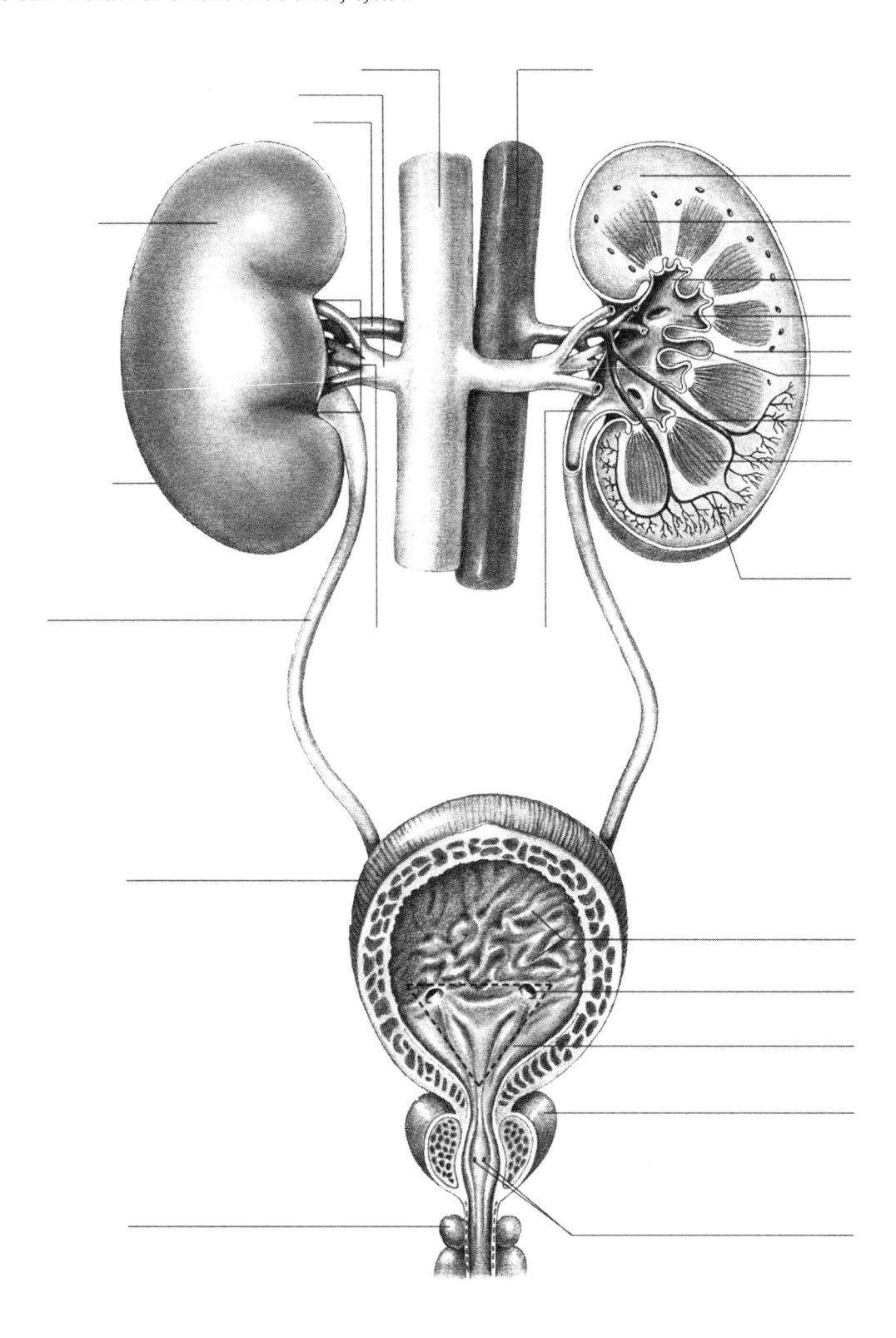

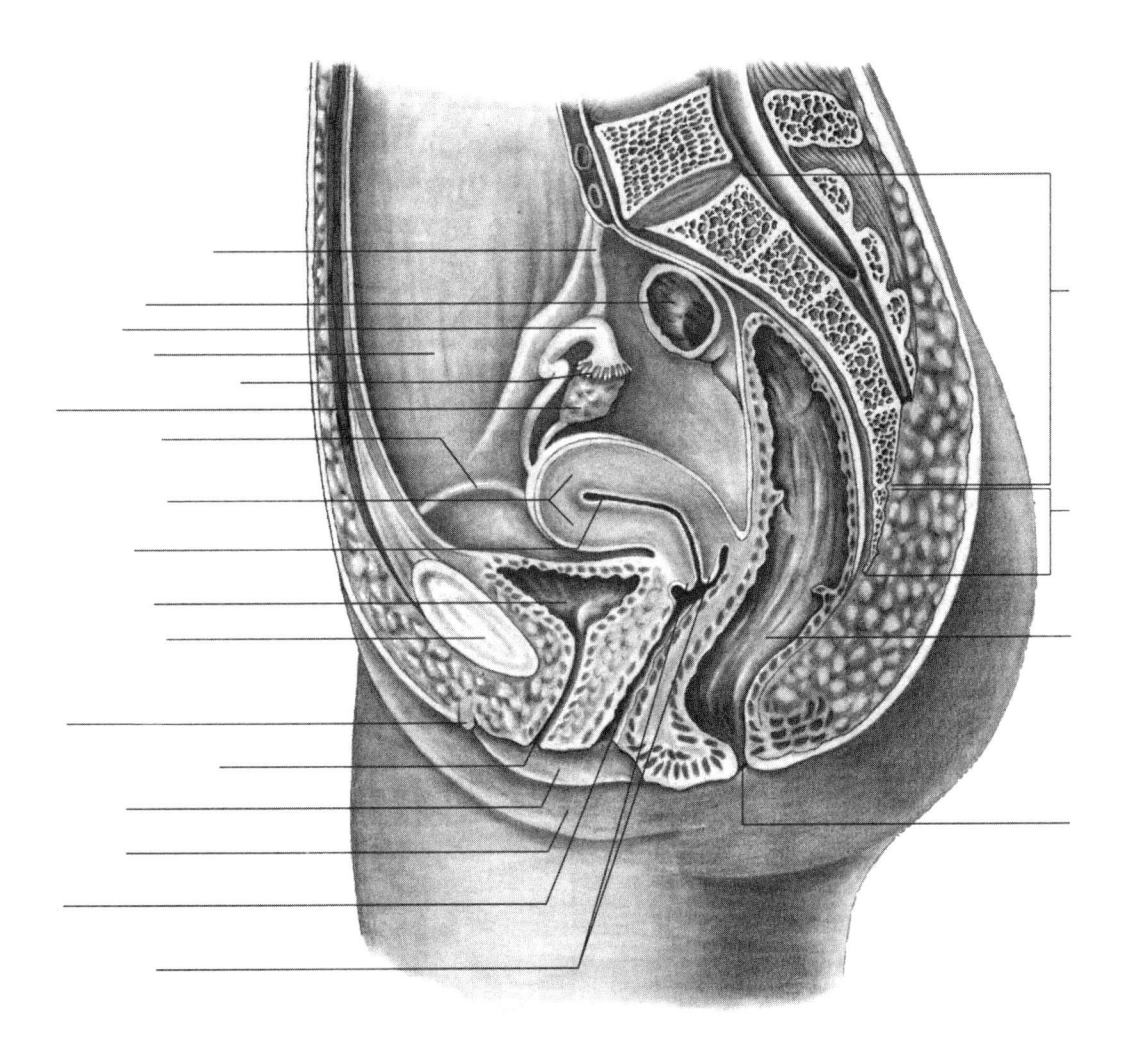

Figure 14.1 Midsagittal section of female pelvis showing internal organs and external genitalia

Figure 14.4 Midsagittal section of male pelvis showing internal organs and external genitalia

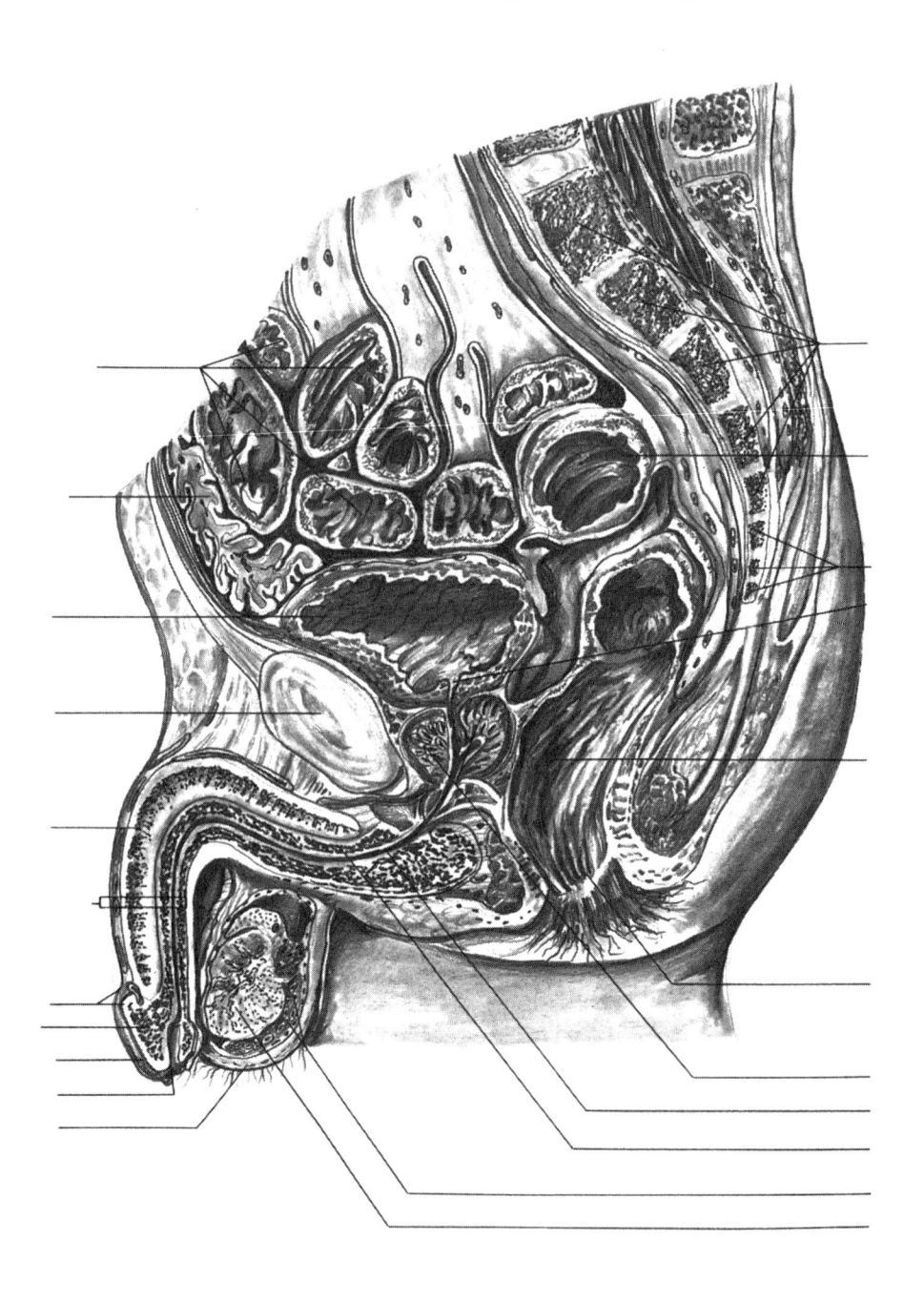